The Panizzi Lectures
1995

Previously published Panizzi Lectures

BIBLIOGRAPHY AND THE SOCIOLOGY OF TEXTS
by D. F. McKenzie (*1985*)

ENGLISH MONARCHS AND THEIR BOOKS
by T. A. Birrell (*1986*)

A NATIONAL LIBRARY IN THEORY AND IN PRACTICE
by K. W. Humphreys (*1987*)

DAPHNIS AND CHLOE:
THE MARKETS AND METAMORPHOSES OF AN UNKNOWN BESTSELLER
by Giles Barber (*1988*)

THE DUTCH AND THEIR BOOKS IN THE MANUSCRIPT AGE
by J. P. Gumbert (*1989*)

ERASMUS, COLET AND MORE:
THE EARLY TUDOR HUMANISTS AND THEIR BOOKS
by J. B. Trapp (*1990*)

THE ENGLISH BOOK IN EIGHTEENTH-CENTURY GERMANY
by Bernhard Fabian (*1991*)

HEBREW MANUSCRIPTS OF EAST AND WEST:
TOWARDS A COMPARATIVE CODICOLOGY
by Malachi Beit-Arié (*1992*)

THE MANUSCRIPT LIBRARY OF SIR ROBERT COTTON
by Colin Tite (*1993*)

MUSIC, PRINT AND CULTURE IN EARLY SIXTEENTH-CENTURY ITALY
by Iain Fenlon (*1994*)

The Panizzi Lectures
1995

Maps as Prints in the Italian Renaissance
Makers, Distributors & Consumers

DAVID WOODWARD

THE BRITISH LIBRARY

First published 1996 by
The British Library
Great Russell Street
London WC1B 3DG

Cataloguing in Publication Data
A catalogue record for this title is
available from The British Library

ISBN 0 7123 4502 7

Designed by John Mitchell
Typeset by Bexhill Phototypesetters, Bexhill-on-Sea
Printed in England by Henry Ling (Printers) Ltd,
Dorchester, Dorset

Contents

Preface

The research for these lectures started with a John Simon Guggenheim Memorial Fellowship in 1977–78. I am most grateful to the Trustees of the Panizzi Foundation for making it possible to summarise my research in these lectures, and to Sarah Tyacke for nominating me to the Trustees in 1990. Elaine Paintin and Karen Cook provided much appreciated help and support in arranging the lectures and exhibition. The British Academy, The Hermon Dunlap Smith Center for the History of Cartography at The Newberry Library, and The Graduate School of the University of Wisconsin–Madison have supported me with fellowships.

Special thanks are due to Karen Cook, Catherine Delano Smith, and Jude Leimer, who made detailed editorial suggestions, and to Corradino Astengo, Francesca Consagra, Conor Fahy, Mary Pedley and Douglas Sims, who kindly read through drafts and offered most helpful comments. Michael Barrett provided bibliographical help supported by the Graduate School of the University of Wisconsin. The University of Wisconsin Cartographic Laboratory drafted the line illustrations in their final form.

Several people have provided valued input into my research on this subject over the years and I am happy to acknowledge them here: David Acton, Nick Adams, Richard Arkway, Luigi Balsamo, Peter Barber, Suzanne Boorsch, Tony Campbell, Susan Danforth, Ed Dahl, Rik De Ghein, Albert Ganado, Paul Gehl,

Gail Geiger, Antony V. Griffiths, Fritz Hellwig, Caroline Karpinski, Robert W. Karrow, Jr., Michael Komanecky, David Landau, Evelyn Lincoln, Carla Clivio Marzoli, Maria Cristina Misiti, Kenneth Nebenzahl, Laurie Nussdorfer, Lucia Nuti, Peter W. Parshall, Monique Pelletier, Armando Petrucci, Franca Nardelli Petrucci, Kiky Polites, Fabia Borroni Salvadori, Alessandro Scafi, Günter Schilder, Domenico Sella, Dirk De Vries, James M. Wells, and Gian Lodovico Masetti Zannini. None of course is responsible for any of my failings.

I must gratefully acknowledge the constant support and understanding of Rosalind, Jenny, and Justin, as well as the staff of the History of Cartography Project, all of whom have shown great patience during what has sometimes been a difficult and exacting task of synthesis.

In his introduction to the Lectures in November 1995, Sir Anthony Kenny kindly dedicated them to Helen Wallis, and it is also my privilege to dedicate this publication to the memory of that wonderful resource of the British Library, indefatigable scholar, and faithful friend.

1

Authors and Engravers

In May 1850, Antonio Panizzi instituted legal action against eight recalcitrant London publishers who had not provided free copies to the British Museum required by the 1842 Copyright Act. Panizzi reported that the first defendant to appear at the Guildhall, 'having begged for mercy, brought a much larger number of books than the firm was known to have published.'[1] Appearing on the same occasion was James Gilbert, who had failed to provide a copy of a map that had been purportedly engraved before 1842 but which was dated 1850. Gilbert complained of Panizzi's actions as 'robbery,' but was fined ten shillings with costs. Gilbert might have more profitably argued that if maps were regarded as engravings, they were not really publications under the Act; The Attorney General stated that bird's-eye views or panoramas were not publications under the Copyright Act, but were an example of 'engraving of pictures' and that 'a collection of engravings with a mere titlepage could not be regarded as a book, unless the plates contained letterpress which conveyed information independent of the engravings.'[2] It is perhaps for this reason that historians of printed books have tended to regard printed maps as eccentric to the mainstream history of book publishing.[3] Yet recent studies of how maps were published and used have revealed hitherto unsuspected commonalities among maps, other printed graphics, and the printed book.[4] The bibliographical and terminological problems in the history of map publishing, and the broader themes of economics, aesthetics, and effects on society, thus deserve special attention.

The Italian map publishers transformed the map from the arcane tools of the navigator, scholar, or administrator into a common article of trade, the geographical print, that became part of everyday life. These lectures examine the roles of authors and engravers, printers and distributors, and patrons and consumers in the map trade that began in Florence in the late fifteenth century, but later developed in Rome and Venice. These last two cities dominated European map printing until they were superseded by the atlas makers in the Low Countries in the last quarter of the sixteenth century.

In fifteenth-century Europe, maps were either highly specialised or highly public. They were not everyday private objects that an ordinary individual purchased, carried about, or kept and displayed at home. It is extremely unlikely that they were used in their most familiar modern role—as wayfinding tools. The Renaissance traveller found his way, as travellers had done for centuries, by asking directions at each settlement along the way.[5]

This did not mean that maps were unknown to the general public in Italy or elsewhere. In the early 1450s, a *barzelletta* circulated in Europe that compared the land of Sicily to the singer's loved-one, Cecilia. The song starts out:

> I have seen the world map, and the navigation charts. But Sicily seems to me to be the most beautiful in this world. I have seen all the islands in the world, but to me the most beautiful island—Cecilia—is not on any map. She came from the other world.[6]

The distinction between a 'world map' and a 'navigation chart' seems to have been evident to those who heard this ballad. World maps were prominently displayed in public places, such as churches or government buildings. Navigation charts of the Mediterranean were practical aids for planning and seafaring. To an Italian humanist scholar, the maps in the classical geographical text of Ptolemy would have been familiar. Academic uses of maps for classical geographical learning and strategic uses for military campaigns were already well known. When the news of the disastrous Venetian defeat in 1509 at Agnadello was received in the capital, the group of senators standing in front of a wall map could understand its strategic significance at once.[7] Such maps were often kept secret to prevent their information falling into the wrong hands. For example, Mantuan artists were refused permission to copy a map in 1506, presumably on security grounds.[8]

The Italian map trade began in Florence at the end of the fifteenth and the beginning of the sixteenth century. Copperplate engraving, essential to the trade, had probably started in that city, where its origin has been convincingly linked to the goldsmith's art. The activity centred around the painter and miniaturist Francesco Rosselli, who engraved and sold maps and prints and who was apparently the first to make a successful independent living from this trade, leaving a large inventory of wood blocks, metal plates, and a stock of impressions that is a key archival document for students of the early print trade.[9]

Despite such a promising start in Florence, the map engraving trade failed to develop there for several reasons. The Italian trade fell into general decline after the shift to the Atlantic economy, when only Venice maintained a commercial presence in the eastern Mediterranean. Florence had been the centre of geographical information in the peninsula during the fifteenth century.[10] The Florentine book trade, so important as an ancillary industry for maps and prints, did not compare to that in Venice and Rome at the time, although books on geography and travel were among the most popular.[11] It may be significant that the only edition of Ptolemy's *Geography* published in Florence was foremost a work of literature, not science, in the humanist tradition—a translation by the priest Francesco Berlinghieri in *terza rima* accompanied by maps that were perhaps printed independently of and before the text (they are printed on different paper).[12]

From the late 1520s onward, Rome and Venice were the two prominent map and print publishing centres in Italy. (Fig. 1)

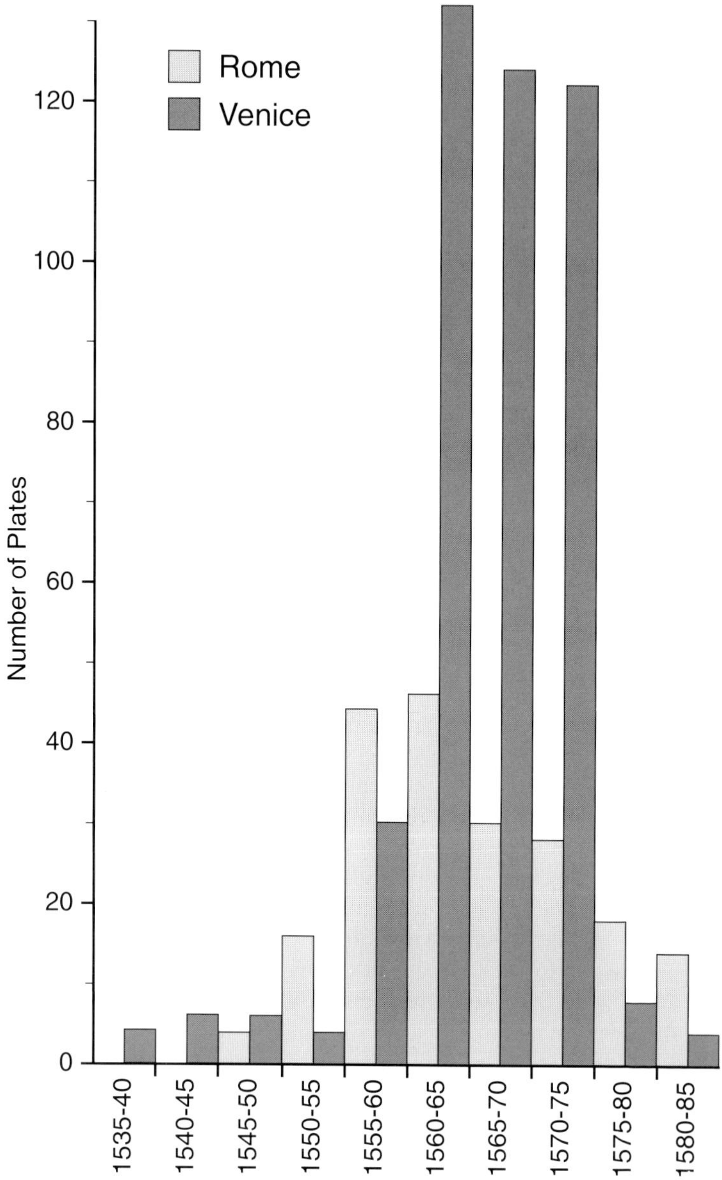

FIG. 1 Numbers of printing plates for maps produced in Venice and Rome by five-year periods from 1535 to 1585. Compiled from data in R. V. Tooley's list of Italian printed maps in *Imago Mundi* 3, (1939) pp. 12–47. Diagram by University of Wisconsin Cartographic Laboratory.

There were differences between them. In Rome, where Antonio Salamanca and Antoine Lafréry were active, the trend in printed maps peaked earlier but lingered longer. The output of map engravers and printers was overall conducted far more intensely in Venice. Five or six hundred copper plates for maps were probably in active use in Venice and perhaps half that number in Rome. The peak in Venice for map publishing occurred around 1566, to be followed by a dramatic decline after 1570. By the mid-1560s, the map trade had matured sufficiently that collections of single maps bound as composite atlases could be purchased to order. The output dwindled to almost nothing after the plague of 1575–77 (in which more than 46,000 people are reported to have died), after which Venetian engravers and cartographers could hardly compete with atlas makers in the Low Countries.

★

A print, geographical or otherwise, goes through three stages of production: creation, reproduction, and distribution. A systematic examination of these three processes in these lectures is offered to provide an understanding of the context of the place of maps in Italian Renaissance society. In the creative stage of map production, the extent to which a map is based on firsthand observation is important if the map is to be viewed as a carrier of geographical information. Central to this issue is the concept of the scale of representation. Firsthand observation is likely to result in detailed maps of local areas, cities, and small regions, and such maps provide the raw material for more generalised compilations of countries, continents, or the whole world. These principles have been recognised by cartographers up to the present, but they were first explicitly discussed in the first book of Ptolemy's *Geography* introduced into Western thought in the early fifteenth century.[13] In addition to a difference in scale, Ptolemy made a fundamental distinction in approach to the two types of graphic description, chorography and geography. Chorography was the work of the artist or painter; it was qualitative and unmeasured. Geography, on the other hand, required mathematics, specifically geometry, to compile maps based on a framework of longitude and latitude. Implicit in Ptolemy's scheme was the overarching study of cosmography, the application of the cataloguing of star positions and their projection in astronomical cartography to the terrestrial world, which he had made explicit in his *Almagest*.[14] A fourth term was added in the Renaissance period: topography. The distinctions were perennially made in geographical treatises throughout the sixteenth century. Thus Giacomo Gastaldi, in the *opuscolo* to accompany his large world map of 1561, states:

> The Ancients found it necessary to describe this sphere with the degrees of the division of the heavens, by way of longitude and latitude, and with lines called parallels and meridians, with the divisions of the frigid and temperate zones and similarly the torrid zone. This general description of the heavens, as with the earth and oceans, the Ancients called Cosmography. Others described in more detail the parts of the earth's oceans, with the divisions of the provinces, kingdoms, regions, and described the cities, castles, mountains, rivers, lakes, seas, gulfs, ports, headlands, and islands, and such description they called Geography. Others described more minutely the parts of that earth and sea as with the navigation charts, and the particular description of territories of cities, with the streets, castles, and villas, rivers, torrents, streams, canals, woods, mountains, valleys between the mountains, countryside, passes and boundaries, seas, gulfs, ports, headlands, bridges, coasts, inlets, reefs, shoals, and other details. Such a description they called Chorography. And the even more detailed description of the site of a fortress, and the plan of that fortress, and private property–houses, *palazzi*, and other similar things they put in Topography.[15]

Typically, Gastaldi's categories of geography, chorography, and topography overlap. For example, the description of castles, mountains, rivers, gulfs, ports, and headlands seem to be included in the domains of both geography and chorography. Gastaldi and other writers were probably concerned only in defining the various terms roughly and, occasionally, in assigning status to the various approaches. Francesco Ferretti suggested that all were important to men who manage affairs of state, but graded chorography, geography, and cosmography respectively as 'Positivo, Comparativo, & Superlativo.'[16]

CHOROGRAPHY

The dictionary of the Accademia della Crusca, the sixteenth-century academy concerned with maintaining the purity of the Italian language, defined chorography as 'An exact description, whether drawn or written, of a given country, province, or territory with respect to its principal partitions, and thus somewhat less detailed than topography.'[17] In 1577, Ignazio Danti wrote that both a geometer and a painter were able to produce chorographies, but called his own measured view of Florence a *topographia*.[18] *Topographia* is also found in the title of a map of Italy.[19] And the term *chorographia* was used also for purportedly *measured* maps of political entities as different as the grand duchy of Tuscany or the country of Greece. Chorographies provided the map archetypes for printed maps of particular regions. By the late Renaissance, compilers of geographies had created a set of authorities for regional maps of Italy, maps whose fame was entrenched through the medium of print. Thus, Girolamo Bell'Armato was the name to know for a map of Tuscany, and he is but one of many such 'authorities' cited by Ortelius in his *Catalogus Auctorum*.[20] Others included Gregorio Amaseo (Friuli); Bonaventura Castiglioni (Milan area); Pietro Coppo (Istria); Agostino Giustiani (Corsica); Pirro Ligorio (Rome); Domenico Macaneo (Lake Maggiore); and Giovan Giorgio Settala (Milan).

For example, two main strands of authority can be distinguished in the regional map of Tuscany: the tradition of the illuminators or miniaturist painters and the measured representations of the military engineers.[21] (Fig.2) Piero Del Massaio represents the illuminator, or 'painter-cartographer' tradition of fifteenth-century Italy, one of the roots of the professional mapmaker's trade. Del Massaio's chorographies of Tuscany appear as 'modern' maps in three manuscripts of Ptolemy's *Geography*.[22] His maps reveal an excellent grasp of the region's geography, and a particularly full understanding of the river and mountain systems.

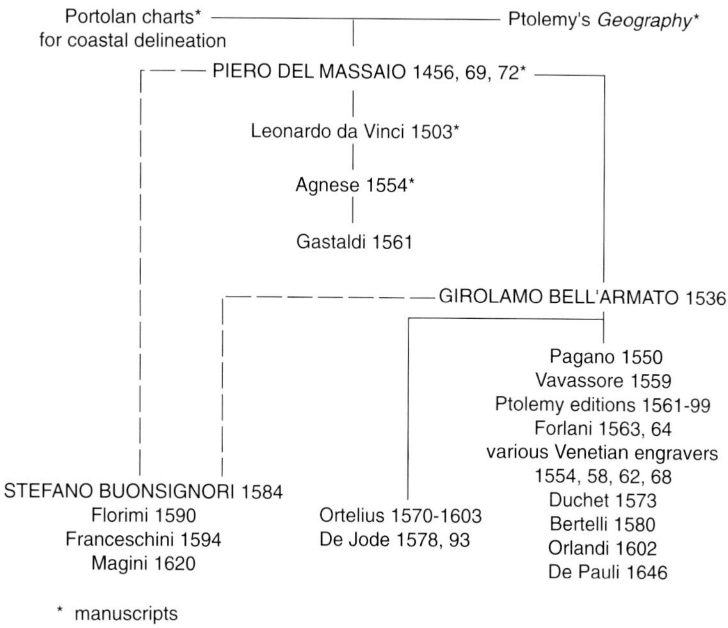

FIG. 2 Simplified stemma or lineage for chorographic maps of Tuscany in the sixteenth century. Diagram by University of Wisconsin Cartographic Laboratory.

It was in non-specialist workshops including graphic artists of all kinds, be they painters, miniaturists, or illuminators, that such maps were produced. The tradition can be traced back to the landscape *vedutismo* in fourteenth- and fifteenth-century paintings by such artists as Ambrogio Lorenzetti and Benozzo Gozzoli.[23] The transition from painted view to map brought not only a change in angle of view but also addition of place-names, transforming the image from a generic to a site-specific representation. Del Massaio's image of Tuscany took the coastline from portolan charts, but it also obviously relied on local knowledge for the interior. His map provided a model for Leonardo da Vinci's manuscript representation of Etruria (1503), and its influence is apparent nearly half a century later in the maps of Italy by Battista Agnese (1554) and Giacomo Gastaldi (1561).[24]

The second thread in the fifteenth century involved the large-scale measured representations of the architect-engineer, which fell into the category of Topography. The tradition of field measurement derived from the Roman *agrimensori* had survived in the Middle Ages. Principles of practical geometry were contributed by the Pisan mathematician Leonardo Fibonacci (thirteenth century) and the astrologer and geometer Paolo Dagomari (fourteenth century). In many Italian cities magistratures had controlled public works since the fourteenth century.[25] In the fifteenth century, the architect-engineers took on administrative roles in large civil and military engineering projects–acqueducts, hydraulic canalisation, and fortifications. The treatises of Mariano di Jacopo and Francesco di Giorgio are the standard examples.[26] The skills of the engineer and the map-maker overlapped. The engineer need not only to be a good draftsman but a good communicator, able to explain his clients' projects with the aid of maps.[27] There are also many hydrographical drawings in the official government archives.

In parallel with the theoretical work on practical geometry there developed the manufacture of topographical instruments and the writing of manuals to explain their use. Indeed, during the sixteenth century, the writing of such treatises became a fashionable pursuit of the gentry with a scholarly mathematical bent. Makers of precision instruments of all kinds included topographic or surveying instruments among their wares; probably the best known in fifteenth-century Italy was Lorenzo della Volpaia (1446–1512) head of the celebrated family of Florentine clockmakers originally from Chianti. One of his sons, Eufrosino, compiled the archetypal map of the territory of Rome in 1547.[28]

The professionalism that characterised the fine arts of sixteenth-century Italy did not extend however to the engineers. The Accademia del Disegno, the brainchild of Giorgio Vasari, catered to the professional needs of painters, sculptors, and even engravers and architects. There were no comparable professional orders of *periti agrimensori, geometri, architetti*, or *ingegneri* in Florence and other Tuscan cities (at least until the eighteenth

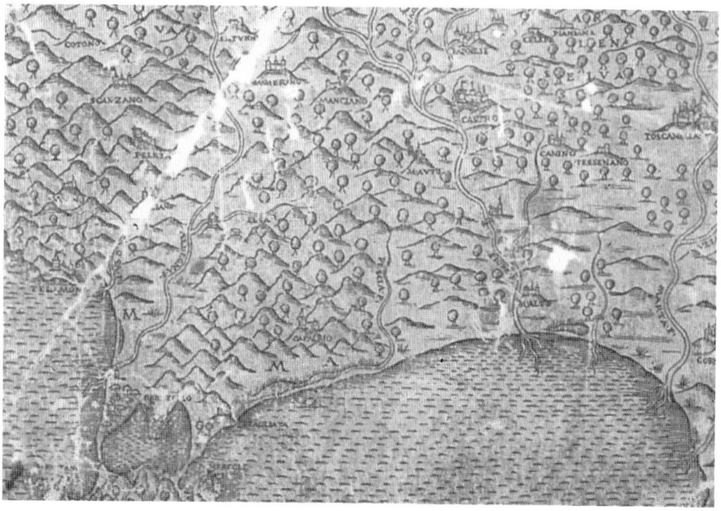

FIG. 3 Detail of the Maremma in Tuscany from Girolamo Bellarmato, *Chorographia Tusciae*, Rome 1536, Florence, Archivio di Stato, Carte nautiche, 13.

century). Education for these trades tended to be private, provided by the technical bureaucracy, the universities, or within the family.[29]

The transition from the cartographic tradition of the painter-miniaturist to the measured observations of a military engineer can be traced in Girolamo Bell'Armato's *Chorographia Tusciae*, published as a woodcut in Rome in 1536. Only one impression survives, and it is in poor condition.[30] The Sienese military engineer indicates in his long dedication to Valerio Orsini that the map is based on direct observations, notations, and measurements taken while travelling on horseback over the greater part of the area. The detailed map pays special attention to the populations of settlements, information useful for military billeting and conscription. The size of the town symbols is based on

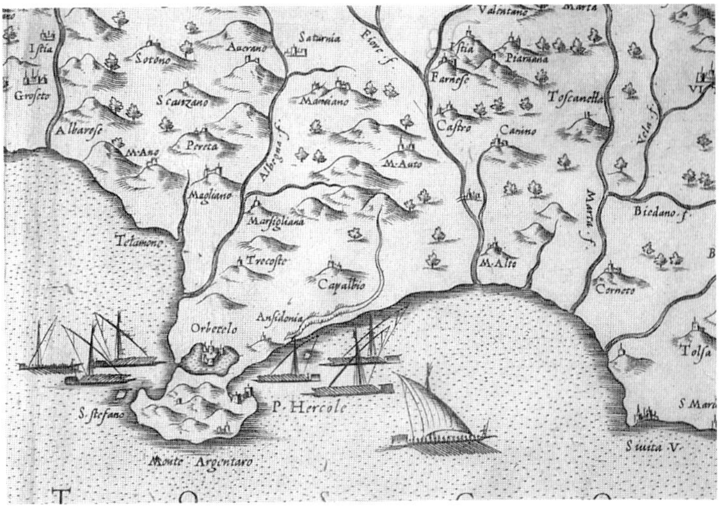

FIG.4 Detail of the Maremma in Tuscany from *Italiae partis Tusciae*, an undated copper engraved reduction of Bell'Armato's map published by Antonio Salamanca, Newberry Library, Chicago Map 4F332.

number of inhabitants: the smallest represents 40-60 hearths, while the others are respectively larger. (Fig.3)

Bell'Armato's map—and the woodcut copies of it made by Matteo Pagano and Gioan Andrea Vavassore in the 1550s—formed the model for the popular copperplate maps of Tuscany sold by the printsellers of Rome and Venice in the 1560s and 70s. Versions appeared by engravers Giulio Sanuto and Paolo Forlani in Venice, Venetian publishers Ferrando Bertelli and Donato Bertelli, and Roman publishers Antonio Salamanca, Claude Duchet, Giovanni Orlandi, and Francesco de Pauli. (Fig.4) Derivations of the map had appeared in nine Venetian editions of Ptolemy's *Geography* by the end of the century. The influence of Bell'Armato's map also spread beyond Italy. Acknowledged reductions appeared in all editions of Ortelius's *Theatrum*,

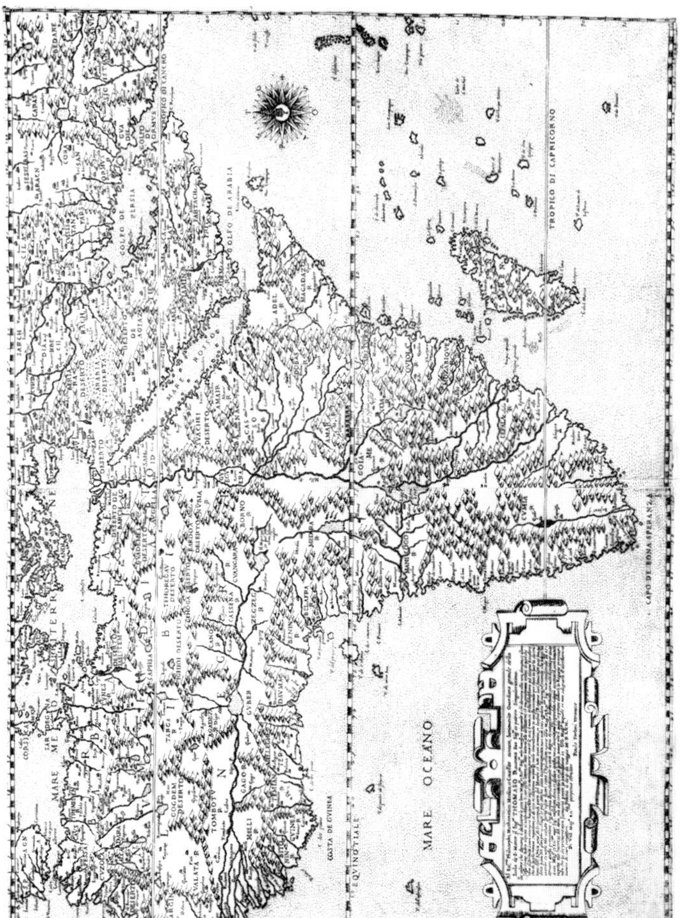

FIG. 5 *Descrittione dell'Africa una delle principali parte del mondo co[n] tutti i suoi termini* . . . Venice, Paolo Forlani, 1562. Newberry Library, Chicago, Map 4F393.

Parergon, and *Epitome*, as well as the *Tabulae Geographicae* of Mercator and the two editions of de Jode's *Speculum*.

GEOGRAPHY

In sixteenth-century Italy, the meaning of *geografia* implied not only a map or description of a larger area than chorography and topography, but also a mathematical methodology using projections and scales. Inconsistency in the use of the term continued throughout the century. While Ptolemy regarded the term as designating a description of the inhabited world—that is the *oikoumene* of the Greeks—the root of *geografia* in fact derives from *gē* meaning the whole world. As knowledge of the inhabited or known world was gradually extended to include the entire globe through the cumulative process of geographical discovery, the Ptolemaic distinction became lost. Indeed, the word was rarely used in the title to indicate a map of the world published in sixteenth-century Europe. Although Antoine Lafréry's title page, which he offered his clients to grace their atlases, included the word *Geografia*, it was clearly intended to refer to regional as well as world maps.[31] Thus the word 'geography' also appears in the titles of maps of parts of Europe, Asia Minor, Africa (Fig.5), Cyprus, Italy, Greece, or the Kingdom of Poland.[32] In short, in sixteenth-century Italy, the term appears to have applied to representations of regions larger than provinces that had some inherent political or physical coherence.

COSMOGRAPHY

The mapping of the entire world was the task of the cosmographer who used a framework of celestially derived circles—parallels and meridians—as a backdrop for explaining the order and creation of the world in metaphysical terms. Cosmography was thus not only a literary *Wunderkammer* of the earth's curiosities; its task was also to show how these strange facts fitted into the broader scheme of things as created by God. Cosmography was universal geography endowed with both the harmony of mathematics and the structure of astronomy; poles, tropics, parallels, and meridians forming a net to capture the world's knowledge. Cosmography thus called on the skills of the geometrician, the geographer, and astrologer, as is shown by the accoutrements in the portrait. (Fig.6) A court-appointed cosmographer could

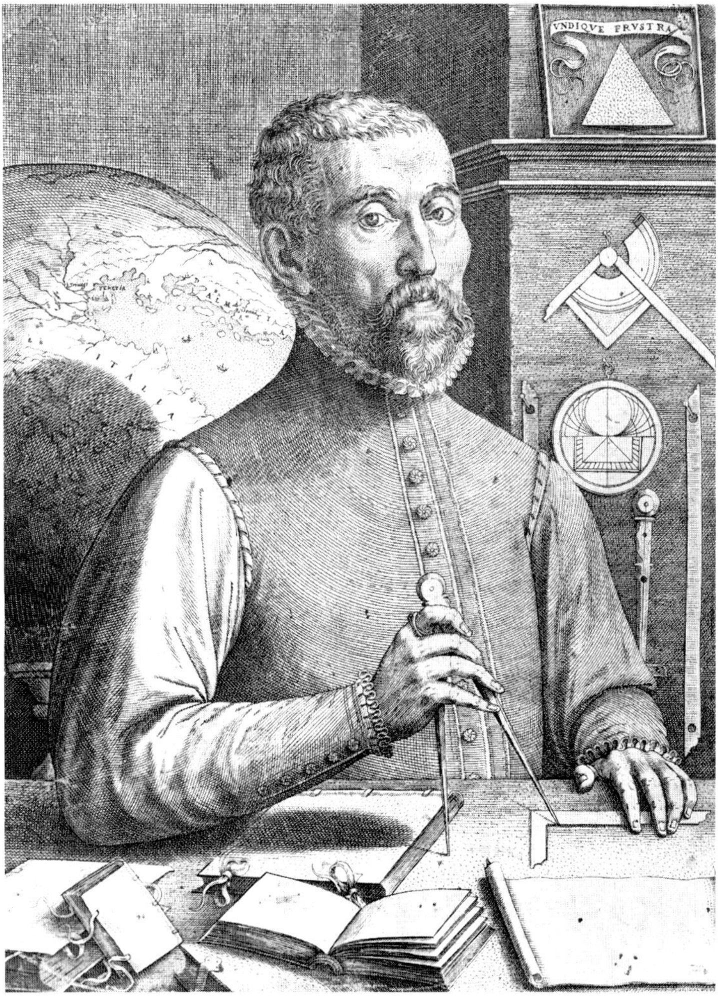

FIG.6 Martino Rota, [Portrait of a Cosmographer], ca. 1570–80. The subject of the portrait is unknown, but, since the three places shown on the globe in the background are Sebenico, Venice, and Treviso, and since Rota was born in Sebenico and worked in Venice, we assume the cosmographer was associated with Treviso. Vienna, Albertina.

reflect the monarch's 'concern with the meanings of space and time which fed into political, religious and historical debate, affecting both humanist speculation and technical practice.'[33]

The notion of authority was obviously crucial and the reputation of the cosmographer of prime importance. For literary and artistic works in the Middle Ages, the sheer age of the sources (texts or artifacts) provided much of the authority. The identity of the maker was of secondary consequence, since it was in any case usually not known. On the other hand, 'Those texts that we now call scientific—those dealing with cosmology and the heavens, medicine and illnesses, natural sciences and geography—were accepted in the Middle Ages, and accepted as true,' only when marked with the name of their author.'[34]

For printed maps in the fifteenth century, the authorities had been the works of the classical geographers: Claudius Ptolemy, Strabo, Pomponius Mela. More printed maps of the world owed their allegiance to Ptolemy in the first fifty years of map printing than to the new discoveries, even though by the end of that half-century period Magellan's *Vittoria* had sailed round the world.[35] Ptolemy was the archetypal cosmographer, and the codification of what shape the world map was supposed to assume followed his instructions for drawing map projections. With the discovery of the extent of the world's oceans and the positioning of the Americas between Europe and Asia, the image of the world changed dramatically. The Ptolemaic map of a quarter of the world evolved into the map of the whole spherical globe on a flat surface. The sixteenth-century world map became known as a *cosmographia*, and the oval world projection became a pervasive cosmographical icon for modernity, universality, and the integration of heaven and earth. Such a world picture was the Renaissance equivalent to that of the *mappamundi* which had been for the previous several centuries an icon of the main events in Christian history played out in the world.[36]

In a different social and economic context, the icon for the maps of navigators, the *carta da navigare*, developed along with the *cosmographia*. Thus, in the sixteenth century, two totally different kinds of icons for world maps were used side by side. After much confusion, including attempts to show navigators'

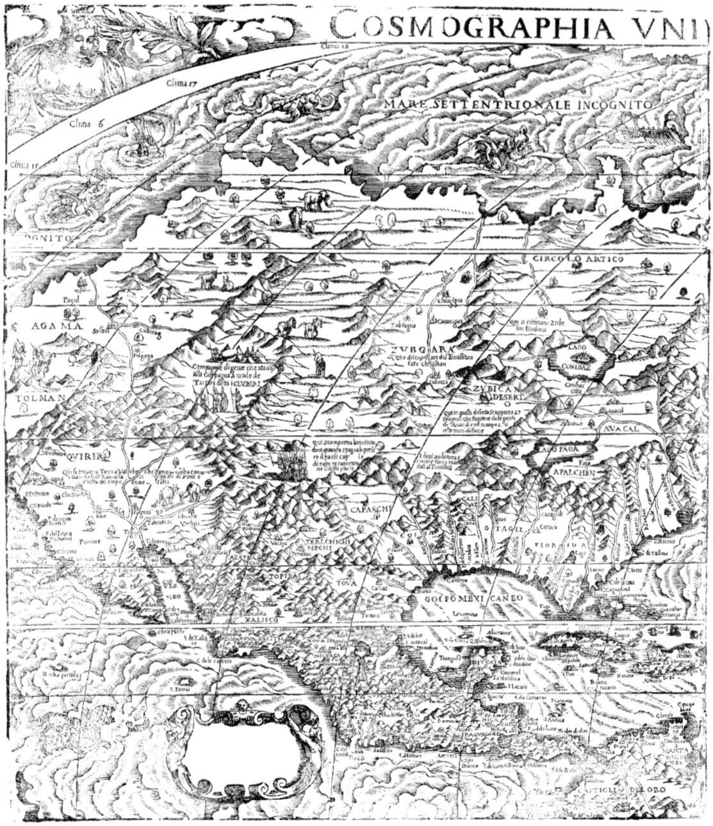

FIG.7 Comparison of details of North America from Giacomo Gastaldi, *Cosmographia universalis*, [Venice, Matteo Pagano, 1561] BL Map Library: Maps C.18.n.1 and Paolo Forlani, *Particolar descrittione di tutte le navigationi del Mondo nuovo . . .* Venice: Simon Pinargenti, 1574. Newberry Library, Chicago, Map 4F49.

rhumb lines on a structure of meridians and parallels, the two traditions converged in Mercator's 1569 projection, which combined the advantages of the cosmographical grid and the plot-

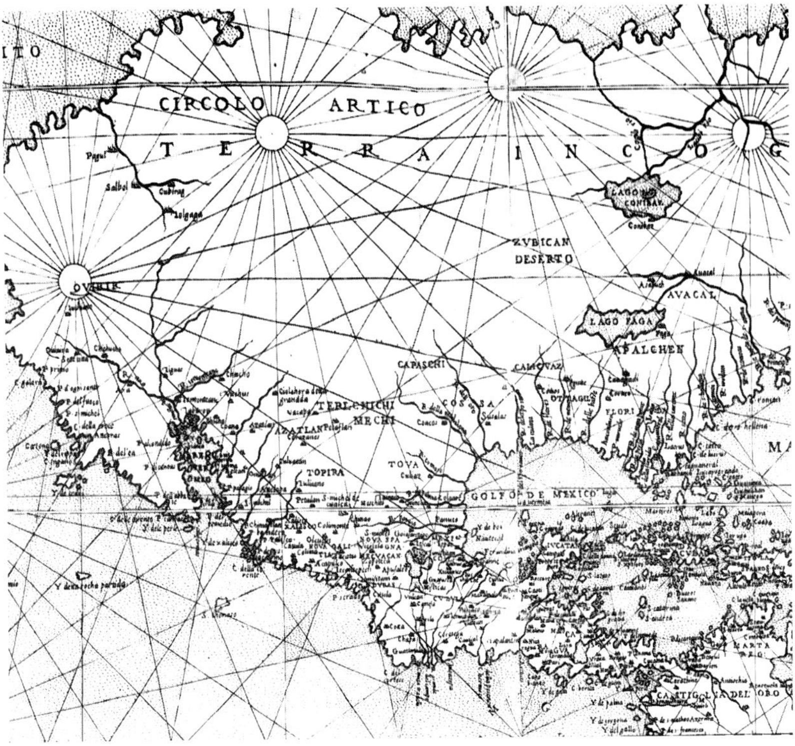

table compass course. As with the use of perspective by artists in the fifteenth century, the use of projections by cosmographers lent professional authenticity to the representation. Indeed, the average engraver and printseller used them far more as symbols of cartographic precision than as meticulous mathematical devices. Mapmakers frequently copied coastlines and cities blindly from one map to another without realising that the projection had changed. An example is provided by Forlani's map of the Americas (1574) which copied the coastlines of the map of the world by Giacomo Gastaldi (1561) exactly while blithely ignoring the change in mathematical structure. (Fig.7)

The Accademia Veneziana epitomised the principles of universality and modernity in cosmography. Nicknamed the Accademia della Fama, because it took as its publishing imprint the image of the goddess of Fame, the Accademia was one of dozens that proliferated in Italy in the latter half of the sixteenth century.[37] The academy's scope was to disseminate science with its own publications, purge the errors from works of all subjects, enrich books with annotations and discussions and translate them in various languages, and finally—like the longer lived Accademia della Crusca in Florence—to promote the Italian language (although this aim had its detractors who feared that the disappearance of Latin would impede the progress of knowledge). The overriding purpose of the Accademia Veneziana was to preserve the political health of Venice, based on the philosophy that 'the observation of divine religion and of the laws, accompanied by the study and love of letters, gives us hope of a secure and enduring state.'[38]

Conceived in 1556 by Federigo Badoer, the academy closed in 1561, suspected by the Inquisition and accused of financial mismanagement.[39] Its members had met 'every day at Badoer's house at the appointed hour to hear lectures and have discussions on every subject.'[40] They included some to whom printed maps had been dedicated: Agostino Valier (Bishop of Verona), Luigi Balbi; those associated as mapmakers such as Alessandro Zorzi, Livio Sanuto; as well as Giacomo Gastaldi. Giuseppe Moleto and Girolamo Ruscelli, respectively the editor and publisher of the 1561 Venetian edition of Ptolemy's *Geography*, were also members. Historians included Paolo Ramusio and Francesco Sansovino.

One of the sections of the Accademia Veneziana was called the *Consiglio delle Scientie*, which was subdivided into four *stanze*, including mathematics. Mathematics included cosmography. In a letter to Andrea Lippomano (another dedicatee on a map) in 1549, Badoer stressed: 'Learning is essential to the prospective statesman. He ought to study cosmography, to learn foreign usages, to pursue Latin and Greek so that he will know rhetoric and moral and natural philosophy, and all those sciences which raise our minds from the earth to heaven.'[41] Fifteen out of the

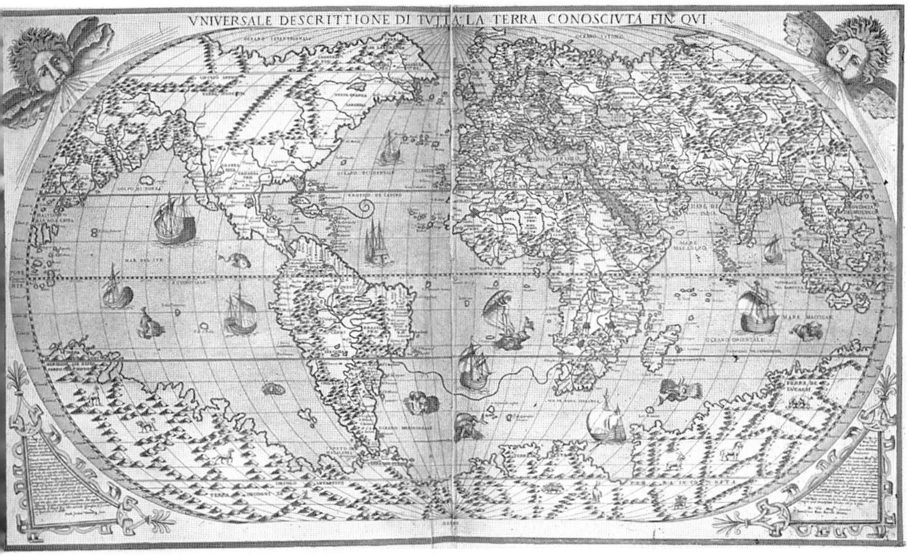

FIG.8 World map at front of composite atlas showing—with a line drawn on the map—the order of plates in the atlas. Paolo Forlani, *UNIVERSALE DESCRITTIONE DI TUTTA LA TERRA CONOSCIUTA FIN QUI*, Venice, Ferdinando Bertelli, 1565. Paris, Bibliothèque de l'Arsenal.

103 books it published were of geographical or cosmographical content, eight classical and seven modern.

The primary Italian authority for the 'universal and modern' maps of the world was 'that excellent cosmographer Giacomo Gastaldi from Piedmont', as he is described in several titles of maps based on his work. Gastaldi's career, strangely obscure for such a celebrated personality during his lifetime—we do not even know the date of his birth—can be sketchily traced from 1539 until his death in 1566.[42] By 1548, he had provided the maps for the first compact edition of Ptolemy's *Geography*, drawn the maps to appear in the *Voyages* of Giovanni Battista Ramusio, and designed the mural maps in the Shield Room in the Ducal

Palace in Venice. From 1551 until his death, his career as a civil engineer and hydrological consultant to the office charged with the regulation of the coasts and waters of the Venetian lagoon can be traced from documents in the State Archives of Venice. In that period, he also compiled several important and influential maps, including the previously mentioned large woodcut world map of ca. 1561, *Cosmographia universalis*.

The image of the new authors is summed up in a common title for world maps: 'the universal description of the whole world known until now.' (Fig.8) The marketing strategy was to provide the reader with *modern* maps, and the use of *new, recent, true, exact, copious*, and *latest* in many of the map titles is understandably evident.[43]

Another indication of the desire to be modern and independent of classical authority is the common use of vernacular Italian rather than Latin in the map titles and legends, with the routine exceptions of world maps and maps of islands. The former were probably in Latin in order to capitalise on a market wider than the domestic. The routine use of Latin for maps of islands, however, reflects a traditional view that the world's islands were under Papal jurisdiction, as a consequence of several Papal bulls of the 1490s.[44]

THE AUTHOR AS DRAUGHTSMAN

The 'universal man' of the Renaissance with any claim to culture had to be able to draw. In *The Courtier*, Castiglione has the Count Lodovico da Canossa point out that the effective courtier should not neglect drawing and painting, even if 'nowadays it may appear mechanical and hardly suited to a gentleman.' Canossa continues:

> From painting, which is in itself a most worthy and noble art, many useful skills can be derived, and not least for military purposes: thus a knowledge of the art gives one the facility to sketch towns, rivers, bridges, citadels, fortresses and similar things, which otherwise cannot be shown to others even if, with a great deal of effort, the details are memorised.[45]

Michelangelo is quoted in a 1585 Venetian edition of the commentaries of Julius Caesar as saying that it was commonly regarded that a military commander should not only use maps and paintings but should also be able to draw:

> The Italians, at least, say that if the Emperor [Charles V], when he attacked Provence, had ordered a drawing to be made of the course of the River Rhône, he could not have met with losses so severe . . . I assert in fact that the modern general of a great army who has not skill and understanding of the art of painting, and is unable to draw, will be incapable of great prowess.'[46]

As evidence of the prestige, perhaps self-assigned, accruing to the delineator or creator of the image, we find a few examples of the likeness of the delineator within the drawing itself. In the often reproduced view of Florence surrounded with a lock and chain, the draftsman is portrayed drawing the city from a perch in the vicinity of Bellosguardo and Monte Oliveto to the southwest of the town. Despite some attempts to discount the possibility of a single viewpoint for drawing the city, it has been demonstrated using modern topographic maps and aerial photographs that the viewpoint could indeed have existed.[47] The interpretation of this image of the author varies: it can be taken to mean that the view was indeed an imitation taken from life, the idea of *contra factum* (counterfeit) in early prints;[48] but it could also function as a graphic signature or an indication of the standpoint from which the drawing was made. Furthermore, the outline of the walls of Florence one can identify on the drawing is part of a play on the relationship between the representation and the world—a picture within a picture.[49]

THE ENGRAVER

The engraver was the intermediary between the author and printer. To get from the intellectual conception of the author to the dissemination of the final print required several stages, including the preparation of a draft for the engraver, engraving, presswork, and marketing the impressions. In sixteenth-century

Italy, we often find practitioners identifying themselves according to one of these roles and using specialised terms to describe their activity.[50]

A central question is to what extent the engraver could assume, or be assigned, responsibility for the content of the map. Too often, scholars have taken it for granted that the name appearing on a map is the 'author' of its content. Thus, *Il disegno del discoperto della nova Franza* (1566 state) has become widely known as 'the Zaltieri map' because it bears the name of the plate's owner, Bolognino Zaltieri. However, since Zaltieri's name was simply added to the second state, it is evident that he had no hand in its compilation or engraving. Even though we can now ascribe the engraving of this plate to the prolific Venetian map engraver Paolo Forlani, we are no more correct in attributing the geographical content to Forlani, for the evidence of Forlani's output points to a slavishly reproductive engraver dependent on the drawings of others. When Forlani tantalisingly engraves 'recently taken from the newest voyages of the French in that region' on this map, we can more realistically accept that information came from a variety of printed sources.[51]

Archival evidence for the costs of engraving and printing maps in the sixteenth century is sparse. The well-known inventory of the shop of Francesco Rosselli (1527) is by far the fullest source.[52] Most of the scattered information about the cost of copper, the cost and time of engraving a copperplate map or print, and the prices of the products in Italy comes from the seventeenth century.[53] Ironically, more cost information is available for the printing trade in the fifteenth century than for the sixteenth. It is also scattered among various Italian cities citing different monetary units, making it difficult to compare.

Notwithstanding the problems involved in making precise estimates of the cost of doing business in the print and map trades of the sixteenth century, it is useful to suggest broad cost parameters while repeating the usual caveats about such estimates. First, labour costs varied according to place and time: for example, the average daily wages for unskilled and skilled labourers in Florence approximately doubled between 1521 and 1571.[54] In the same period the price of a pound of beef and lamb also approximately a little more than doubled. Inflation was also reflected in the value of the florin against the lira in Florence, which rose from 140 soldi in 1500 to 150 soldi in 1533.[55] De Vries and other economic historians speak of a golden age of labour in the fifteenth century, followed by a disastrous 'long sixteenth century' in which inflation outstripped the rise of wages.[56] For Venice between 1567 and 1599 the average number of days work required for subsistence grew as follows:

	1567	1599
master	20	35
worker	30	50
master's family (3.5 persons)	60	120
worker's family (3.5 persons)	90	170[57]

Second, the economy in Florence, Venice, and Rome was not exclusively based on cash. Although skilled manual workers—engravers or pressmen—in the print shops were normally paid in cash or even a salary if they were much in demand, the low wages of apprentices and unskilled bachelor workers often included lodging.[58] The payments in kind to artists, authors, and other professionals recorded by Francesco Calzolari, Jr., in his account book bear witness to the extraordinary complexity of these transactions (even down to gifts of melons, trout, and marzipan for the Inquisitor).[59] For the shop-owners, and occasionally for employees, their business also involved extending credit. In Rome, for example, Antoine Lafréry negotiated to hold several of Venturino Tramezzino's copper plates as collateral for a loan to cover the expenses of a deal Venturino wished to make but for which he was short of cash. A document of January 1574 describes the use of copper plates as collateral, i.e., 'two copper plates of England, one of Holland; one of Spain; one with "il Settentrione"'. Lafréry paid at least thirty-one scudi for these five plates, pledging not to use them and to sell them back to Tramezzino, which he did a little before the deadline.[60]

Technical information on copper engraving—and the map trade generally—is virtually unknown in the Venetian archives, although this does not mean that none exists. There are copious primary records of the activities of dormouse skinners, but unfortunately few for map engravers.[61] In the absence of primary sources, we are obliged to reconstruct what day-to-day issues might have arisen in the course of a printmaker's work from the output of the engravers themselves. Paolo Forlani makes a good example, as he concentrated almost entirely on map engraving.[62] The recently published engraving manual of Domenico Tempesti helps to fill in the details, for Tempesti, although basing much of his information on the manual of Abraham Bosse, adds details of his own, particularly in the printmaking phase of the operation.[63]

As a sample, take the year 1566, the zenith of Venetian map engraving. This year accounted for about a quarter of Forlani's total output of plates over an eleven year period.[64] During those twelve months, if they were to take the usual religious holidays, he or his assistants needed to engrave about 13 × 13 cm every working day.[65] At this rate, a plate in the *reale* size (about 29.5 × 43.4 cm or 1280 cm^2) would take about eight working days to engrave. A small *octavo* plate might take about one day. Help was certainly necessary, because Forlani ran a print shop as well as an engraving studio.

The cost of copper was considerable. To save money, maps and prints were sometimes engraved on both sides of a piece of copper. The Rosselli inventory mentions such cases. In 1613, we learn that a tinker specialising in weights and measures purchased a copper plate from a shop on the Piazza Pasquino in Rome for 22 baiocchi per *libre* (*libbra*) of copper (.22 scudi) hoping to use it as a platform for one of his scales.[66] In 1669, a copper plate in the imperial size (about 70 × 50 cm) cost 6 scudi.[67] During the early part of the sixteenth century the trade in copper ore issuing mainly from Hungarian mines came largely under the control of the Fugger enterprises in Augsburg, but Italy also imported it from various other places such as the Tirol.[68] As the Renaissance metallurgist Biringuccio tells us in his 1540 treatise on the subject, refined copper came in the form of cakes or ingots that could be hammered into sheets either by a professional coppersmith or, less likely, by the printmaker himself.[69] Around the middle of the century a new method of rolling copper plates came into use that allowed the surface of the copper to be much freer from imperfections than hand beaten material.[70] From the Rosselli inventory, we know that the plates weighed 475 libra (about 161 kg) and some of the plates are described as being *fogli comuni, fogli reali, fogli mezzani* from the sizes of paper they might be expected to print.[71]

The plates were scraped smooth with a knife and cleaned with soft chestnut carbon and pumice stone, taking care to avoid scratching the copper.[72] Nevertheless, scratches can usually be seen, and indeed are sometimes useful to the bibliographer because they signal the amount of wear to which the plate has been subjected. The two states of the Forlani map of North America can be identified on the basis of the extent of the plate scratches alone.[73]

To transfer the information to the plate, the engraver waxed the plate lightly, took a drawing or print to be engraved, varnished it, and laid it face down on the waxed plate, tracing through the main lines to the wax beneath. Once these were in place, the detail could be copied by hand directly onto the plate. Straight lines and hatching were left to assistants. Doodling in the margin was common, ranging from practice marks to an intimate note that would be covered over by overlapping sheets.

Leaving the lettering until last, the engraver sketched it in lightly with a drypoint and then finished it with a burin. Forlani was known for his skill in engraving lettering quickly and effectively in his characteristic style, and such work would not have been entrusted to an apprentice. We can recognise his hand in some place-names added to the bottom right corner of a 1566 map of Switzerland made by Domenico Zenoi. The engraver worked under the window at a long bench covered with cowhide and held the plate on a small leather cushion filled with sand, moving it around as needed.[74] His oiled sharpening stone—the best type of marble from the Levant—was constantly at the ready to sharpen his burin.[75]

Forlani probably commanded prestige as an engraver because he could engrave lettering. Lettering engravers were paid at special rates and were sometimes specifically acknowledged. One receipt dated 1690 records the payment of 6.60 scudi to an unknown 'Sicilian who engraved the letters on the frontispiece.'

Giovanni Giacomo de Rossi sometimes added the name of the letter engraver on his plates, Falda's Plan of Rome (1675), for example, which carries the inscription 'Giorgio Widman intaglio le littere'.[76]

The cost of labour for engraving is again reconstructed only with difficulty from scattered references. In 1635, Giacomo Lauro petitioned the court to establish the identity of the thief who had stolen a copper plate depicting the Senate of Poland and the grand duke of Moscow from the residence of the Bishop of Luck, the Polish ambassador to the Papal States, for whom he had engraved it in 1613. We learn that Lauro spent forty scudi producing the plate and pulled prints from it as a gift for the bishop, who paid Lauro 30 gold scudi for the honor of borrowing the plate and for the dedication copies.[77]

The account book for the manufacture of the *Musaeum* of Francesco Calzolari, Jr. also allows us to draw some conclusions about the average cost of producing copper plates. Calzolari, a Veronese apothecary, published a large illustrated volume describing his family's natural history collection in 1622. The average cost per engraving of 36 of the 45 plates for the *Musaeum* for which we have figures was 11.1 Veronese lire (= about 14.8 Venetian lire) for an average size plate of 208 cm^2.[78] The account book provides specific information about the drawing and engraving of the complex frontispiece view. The artist, Giovanni Battista Bertoni, was paid 36 Veronese lire (48 Venetian lire) for the drawing, while Girolamo Viscardi was paid twenty ducats, or 93 Veronese lire (124 Venetian lire), for cutting the plate.[79] These figures confirm that designers of drawings for plates were generally paid less than their engravers. The painter Rubens 'acknowledged that in book illustration the creative work of the artist had to be less well remunerated than its technical execution.'[80]

A number of other cases of payments to engravers have been documented. Consagra points out that labour costs varied according to medium, complexity and size of image, status of the printmaker, and the extra labour necessary for lettering and backgrounds. Engraving took longer than etching, was regarded as of higher quality, and was thus better remunerated. Northern engravers, who enjoyed a good reputation, could command higher payments than their Italian counterparts, which brought 20–100 scudi per plate.[81] Other controls and incentives were sometimes imposed or provided: Natale Bonifacio agreed to work on a commission with Andrea Bacci, a doctor. To avoid his being distracted, he was obligated to work in the doctor's house, not to make other copies, and to finish the work within 40 or 50 days. The payment was 25 scudi, with a bonus if the work succeeded well.[82]

The frequent anonymity of engravers is a puzzling feature of sixteenth-century Italian engraving. Forlani only signed about eighty of over 144 maps we believe he engraved over eleven years.[83] There was a fairly cavalier attitude about adding to another engraver's work (either by adding new plates or by adding place-names to another engraver's existing plate). Sometimes, if the map was pirated or copied, the engraver's name remained, but such cases were infrequent.

The frequent omission of engravers' names is sometimes frustrating to twentieth-century map and print historians. Tooley even tried to invent an engraver in his bibliography of sixteenth-century Italian maps, claiming that one 'Mugnone' had signed a view of Florence as 'mugnone f' or 'mugnone fecit.'[84] The phantom engraver's name actually refers to the 'Mugnone fiume' a well-known *torrente* that was diverted around the fortifications of the city. The error was copied by Marcel Destombes in his checklist of Italian printed maps in the Bibliothèque Nationale.[85]

THE COPPERPLATE 'MAP AESTHETIC'

We can discern a 'map aesthetic' in copperplate maps that carries over into the broader context of prints. Maps often involve a high degree of line definition to reflect the precision of a boundary or of measured geometry. Bosse described one of the qualities valued in the earliest impressions of prints as the stark contrast of distinct black lines against very white paper; the absence of colour enhanced this aesthetic.[86] A clean background was needed to avoid confusing blobs or imperfections in the plate with geographical information. There is also a technical as well as aesthetic reason for this: fine lines break down quickly during printing. For example, the maps of Rosselli are bold and vigorous in their style, and we see a similar vigor in the little known map work of Girolamo Mocetto, who was responsible for four curious engraved maps in Ambrosius Leo's book on the town of Nola published in Venice in 1514.[87] (Fig.9)

In addition to this boldness of style, another 'mapping effect' in prints concerns the role of *annotation*. Other types of graphics, such as commemorative mosaics, have long been annotated, but maps routinely require names, labels, titles, text, and many of the most informative features of a map are the expository legends scattered around them. Calligraphy and cartography were closely linked, and the compact, efficient, legible, semi-formal hand adopted in the Vatican Chancery formed the basis for the copy books of Arrighi, Tagliente, Palatino, and Franco, the last two of whom also engraved maps.[88] Michael Bury has pointed out that Giulio Sanuto's style of engraving figurative prints suggests 'that he expected his figural print to be used in a manner analogous to the way a map was used. . . The habit of annotation, providing a written commentary on the representation, provides an example of an attitude of mind deriving from his experience as an engraver of maps.'[89]

As in music printing, punches were commonly used in map engraving to avoid the time-consuming engraving of small letters. Giulio Sanuto provides an excellent example, with his maps for his brother Livio's atlas of Africa (1588) and for the maps in various Venetian editions of Ptolemy's *Geography*. In theory, punches could be catalogued and possibly traced through various media, but their role as a means of identification has yet to be fully exploited.[90] The lettering method was exploited extremely early in the history of map printing—for example for the Rome editions of Ptolemy in 1478 and 1490. When new maps were added in 1507, the punches were reused, with the exception of the 'O' punch, which had obviously become lost, so that a punch used to engrave town circles was used instead. (Fig.10)

The choice of whether to use a woodcut, an engraving, or an etching in the sixteenth century had aesthetic and technical considerations. Etching lacked the crispness and effect of precision

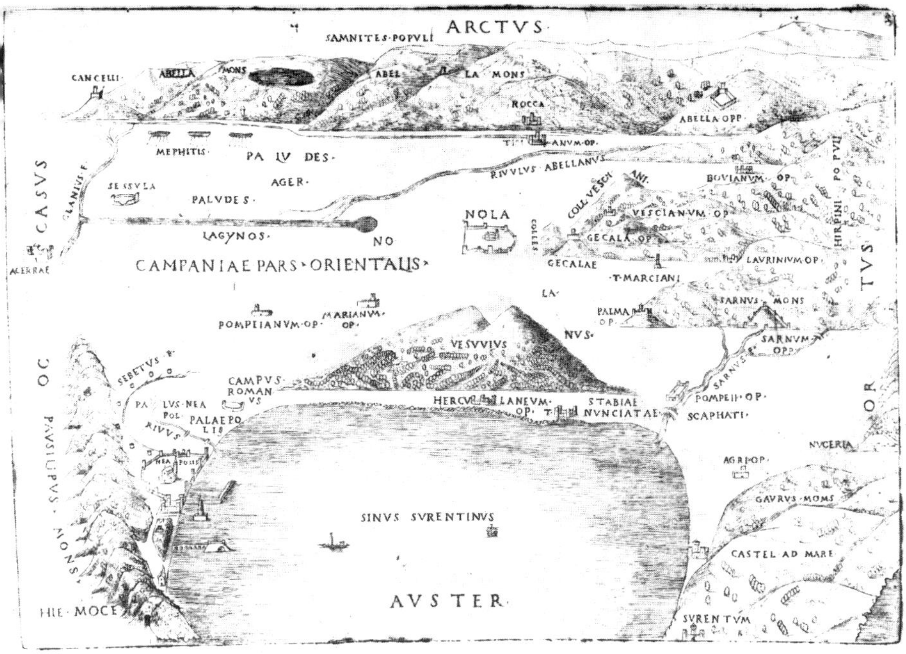

FIG.9 Girolamo Mocetto, [untitled map of environs of Nola], facing p.iiii in: Ambrosius Leo (Nolanus), *De Nola, opusculum distinctum, pulcrum, verum, grave, varium, et utile,* Venice, 1514. British Library, 178.g.14.

given by the burin, but it was cheaper, as we have seen from the difference in wages paid to etchers and engravers: the increased cost of engraving with the burin no doubt contributed to its snob value. Etching had an uneasy start early in the sixteenth century pending suitable mordants for various metals. Iron and steel were first used, presumably linking back to the origins of the etching method for armour and other decorative metalwork.[91] Wolfgang Lazius etched his own plates on iron for a series of maps of Austria in the mid-sixteenth century, but copper etching was not well known for maps or prints in Italy until

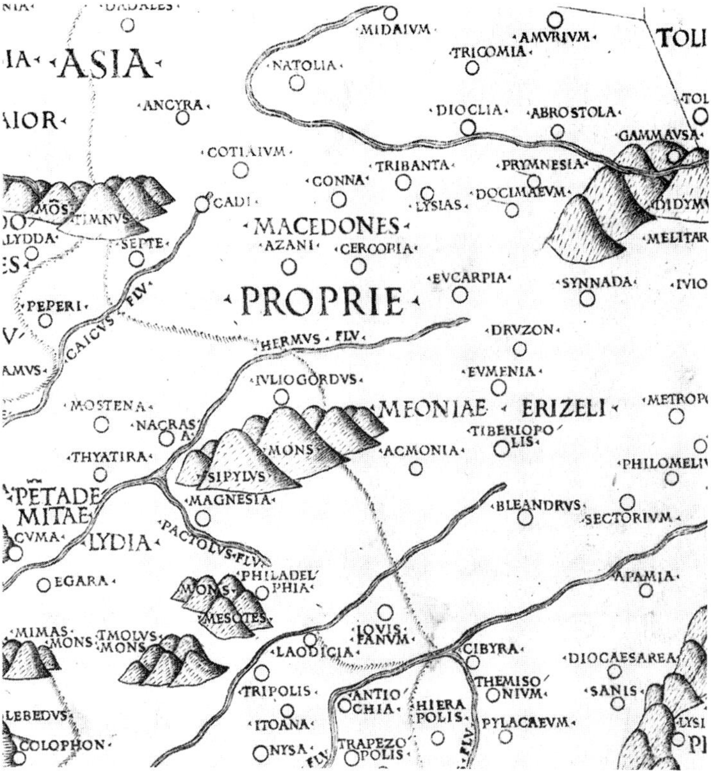

FIG. 10 Details showing substitution of 'O' punch in maps added to the 1490 edition of Ptolemy. Compare the 1490 regional map in Claudius Ptolemaeus, *Cosmographia* Rome, 1490 and Johannes Ruysch, [world map] in Claudius Ptolemaeus, *Cosmographia* Rome, 1507. BL Maps C.1.d.3 and C.1.d.4.

the second half of the century, and was not in widespread use until the seventeenth.[92] The etching needle could not easily achieve the elegance of calligraphic lettering that required stresses and movement to produce subtle changes in the thickness of lines. An etched plate, in which the lines were bitten shallower, wore out more quickly and yielded only about half the number of impressions from an engraved plate.[93] The

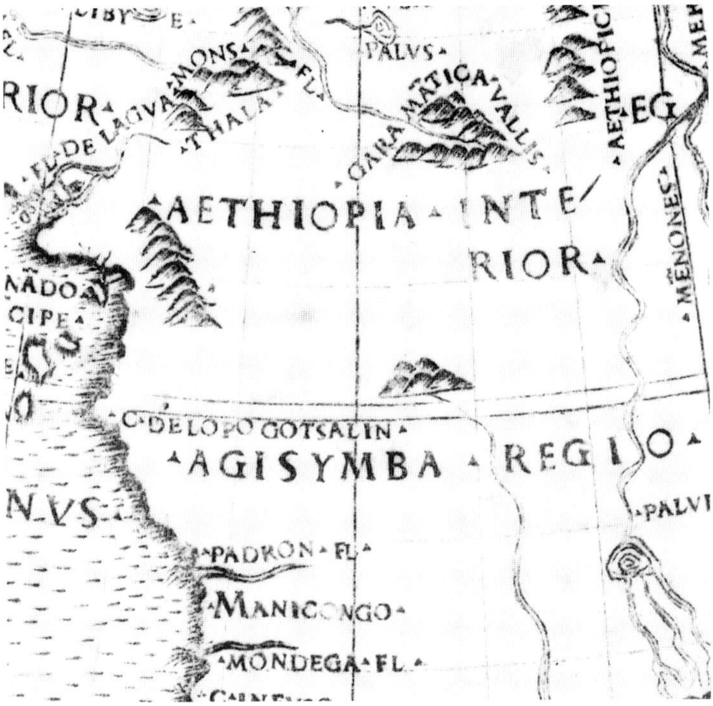

process was messy, and great care had to be taken to avoid scarring the plate with acid. If not carefully cleaned, this scarring and pitting would attract dirt and ink. The etching process was also extremely dangerous: great care had to be taken to avoid damage to the throat and lungs from the nitric acid fumes. Contemporary manuals on armour etching, and modern etchers confirm the extent of the precautions necessary to vent these fumes from the workshop.[94]

The autography of etching made up for its shortcomings. Its stylistic versatility could cater to the increasing demand for a wider range of tonal effects and artistic expression. Thus etching was used for the decorative details of many maps, reserving the burin for the more exacting cartographic information. Forlani apparently used etching early in his career (in the early 1560s), for three of his early maps have rounded lines of consistent width and appear to be at least partly etched, but he quickly abandoned it.[95]

WOODCUT AND COPPERPLATE COMPARED

The generalisation that Italian engravers worked mainly with metal plates and that their counterparts north of the Alps worked in wood is broadly true.[96] In the sixteenth century, 'blockcutters' were more active in Germany, where the single-leaf woodcut became the characteristic genre for printed illustrations of all kinds, and where the links with the book publishing trade (which used woodcut illustrations heavily) had already been strongly established.[97] However, Venetian publishers tended to use blockcutters more than their Roman equivalents in the first half of the sixteenth century; the best-known artisans were Benedetto Bordon, Giovan Andrea di Vavassore, and Matteo Pagano. Reasons for this use of the woodcut may lie not only in the strong cultural affinities of Venice with cities north of the Alps, but also in the early dominance of the major painters in each city who used reproductive prints to help disseminate their work. In Venice, where Titian's influence predominated, the woodcut was used to express the master's art in prints, as in those by Nicolò Boldrini. In Rome, where the copper engravings of Marcantonio Raimondi reproduced the art of Raphael, the taste for the intaglio medium became established earlier.[98]

The rugged style of the woodcut lent itself to large multi-sheet maps and views intended to be viewed at a distance, and for pattern books, broadsides, or other illustrations in which delicacy of line was not the prime consideration. The examples are well known: Jacopo de' Barbari's view of Venice, the view of Florence 'with the chain' derived from Rosselli's copper engraving, and Gastaldi's *Cosmographia universalis* of 1561. Furthermore, for a printer who already owned a common printing press that could produce books from moveable type and who already might have a blockcutter available for book illustrations, the woodcut was the most convenient technique even for single-leaf illustrations.

These single-leaf woodcut maps in sixteenth-century Italy are extremely scarce. Here we can cite, as an example, the corpus of Giovanni Andrea di Vavassore, called Guadagnino.[99] Vavassore began as an artist and blockcutter around 1510–15. He has been confused with at least three other woodcut artists of the period: Zoan Andrea, a Mantuan, who copied many of Mantegna's works; Zoan Andrea, now known as 'the Apocalypse artist'; and 'I. A.', an illustrator of religious works.[100] From 1530 Vavassore is listed as a member of a painters' guild that admitted artists and artisans. He had at least two brothers, Florio and Luigi, both blockcutters. Vavassore's map of Rhodes is dated 1522. His earliest known book decoration is the frontispiece (signed Zovan Andrea de Vavasori) to a work published by Zoppino in 1518.[101] In 1537 his publishing firm was described as 'Giovanni Andrea detto Guadagnino e fratelli di Vavassore'. From 1544–72 the imprint bears only Giovanni Andrea's name and it is during this period that he published an illustrated edition of Ariosto's *Orlando Furioso*.[102] It is difficult to fix the beginnings of Vavassore's cartographic activity, for about half his maps are undated. None of his works survives in more than five or six impressions.[103]

Several factors may account for the low survival rate of woodcuts in Italy. Since the blocks were faster to produce and cheaper than copper plates and the paper less expensive and rarely watermarked, woodcut prints were cheaper for the consumer. Size for size, they cost about a tenth or a twelfth of the price of their copperplate equivalents.[104] Woodcuts were consequently less collectible and more ephemeral; if not bound into composite atlases or books, it is doubtful that any would have survived. Many must have disappeared, leaving no trace either of surviving impressions or documentary evidence, a situation parallel to that of printed books of the period.[105] Serendipidity often accounts for finding these rarities in unusual places. For example, an impression of Vavassore's view of Venice was found folded and bound in a copy of Fabio Calvo's *Antiquae Urbis Romae* (1532) during its restoration following the 1966 Florence flood.[106]

Despite their many advantages, woodcuts were cumbersome

to cut, particularly when a large amount of lettering was involved. As has been widely documented, many ingenious methods for producing lettering on woodcut maps were developed.[107] Publishers even went to the extreme of pasting small labels over their mistakes, as in the 1511 edition of Ptolemy's *Geography*, where the towns of Modena (Mutina) and Bologna (Bononia) had been reversed. (Fig. 11) Fine lines were difficult to carve and innately brittle in the press. Large wood blocks of sufficient fine grain and flatness must have been difficult to obtain, particularly when the usual supply of suitable wood from Asia Minor was cut off during the Turkish wars.

Despite their technical problems, many woodcut maps served as prototype versions of maps that were slavishly copied by the copper engravers. Woodcuts often provided the archetype for regional maps, as with Bell'Armato's map of Tuscany and Leonardo Bufalini's large woodcut map of Rome. Artist-architect Bufalini, who worked as a blockcutter for Antonio Blado, had designed, published, and probably executed, in wood, the oldest of what are now regarded as the great maps of Rome.[108]

Matteo Pagano's work is another excellent example of the use of woodcuts as prototype maps. To judge from his dated works, Pagano was active as blockcutter and publisher from 1538 to 1562 in the Frezzaria 'all'insegna della Fede.'[109] Pagano's association with Giacomo Gastaldi was much stronger than Vavassore's; he probably cut the small woodcut maps (apparently constructed by Gastaldi) that illustrate the collection of accounts of voyages by Giovanni Battista Ramusio, and he published Gastaldi's brief booklet that accompanied Gastaldi's 1561 world map.[110] A 1565 edition of this booklet was published by Francesco de Tomaso di Salò, who took over his establishment after his death and over whose imprint Pagano's fine reduction of Barbari's view of Venice was published (Fig. 12). As with Vavassore, Pagano's eighteen surviving woodcut maps are now extremely rare. In no case does any map survive in more than three impressions. Although the output of the two engravers was similar in both content and style, Pagano apparently engraved for more authors of prototype or archetypal maps, whereas Vavassore seems to have been more of a copyist. For example, Pagano's map of Piedmont served as a

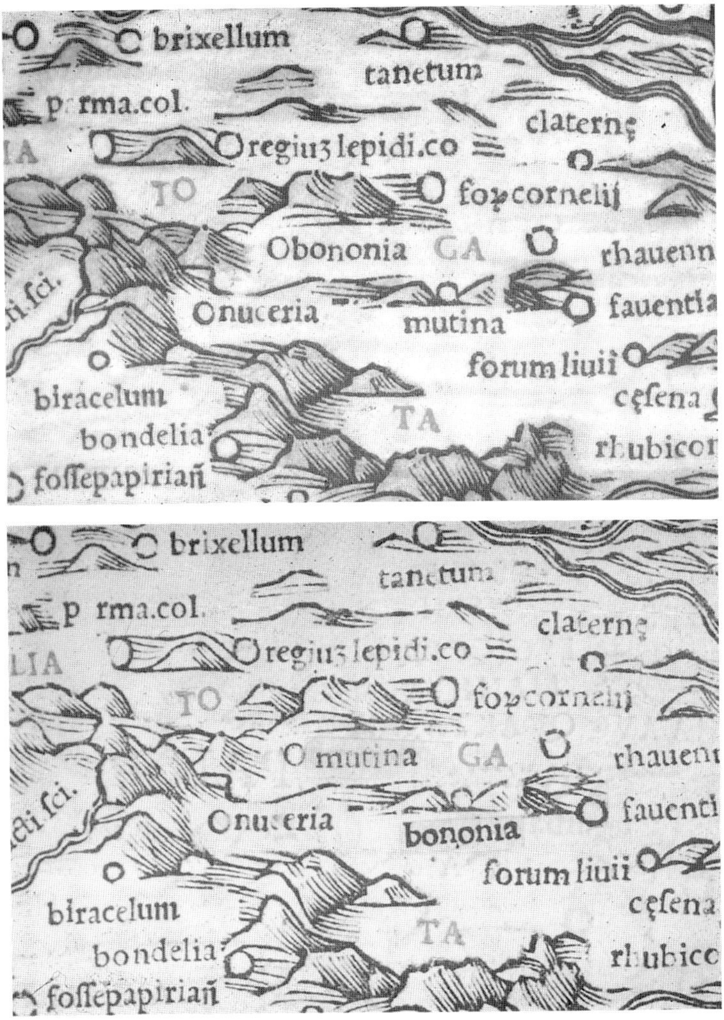

FIG. 11 Comparison of two variants of the 1511 edition of Ptolemy's *Cosmographia*. In the earlier, vellum version (BL Map Library GR 8176) the location of the towns Mutina and Bononia (Modena and Bologna) has been reversed. In the later version (BL Map Library C.1.d.7), small slips of paper were added to correct the error.

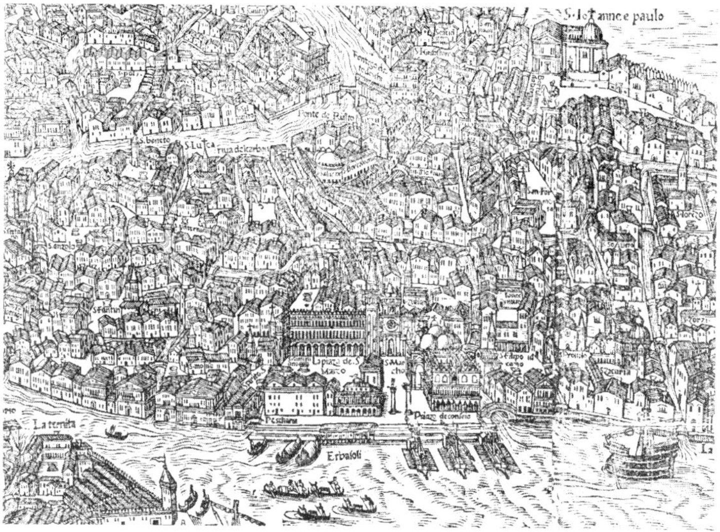

FIG. 12 Detail of the map and printselling areas of the Merzaria and Frezzaria in Venice between San Marco and the Rialto from Matteo Pagano's *Venetia*, Venice, Francesco e Tommaso da Salò, 1567. BL Map Library *22670.(11.)

model for that of Hieronymus Cock (Antwerp 1552) who also made a few corrections using Gastaldi's representation of 1548 among other sources.[111] (Fig. 13)

Only a handful of surviving maps were engraved on copper in Italy from 1508 to the Gastaldi map of Spain in 1544: Girolamo Moceto's plans of Nola (first decade of sixteenth century);[112] a local map of the district between Milan and Adda in *Decretum super flumine Abduae reddendo navigabili Mediolanum usque* (Milan, 1520),[113] and a world map by Vespucci (1524).[114] The gap in the use of copperplate engraving for maps printed in Italy between 1508 and the 1540s poses a puzzle. In the first four decades of map printing, copper engraving had apparently been successful, considering its use for the Bologna (1477), Florence (1480), and four Roman editions of Ptolemy's *Geography* (1478, 1490, 1507,

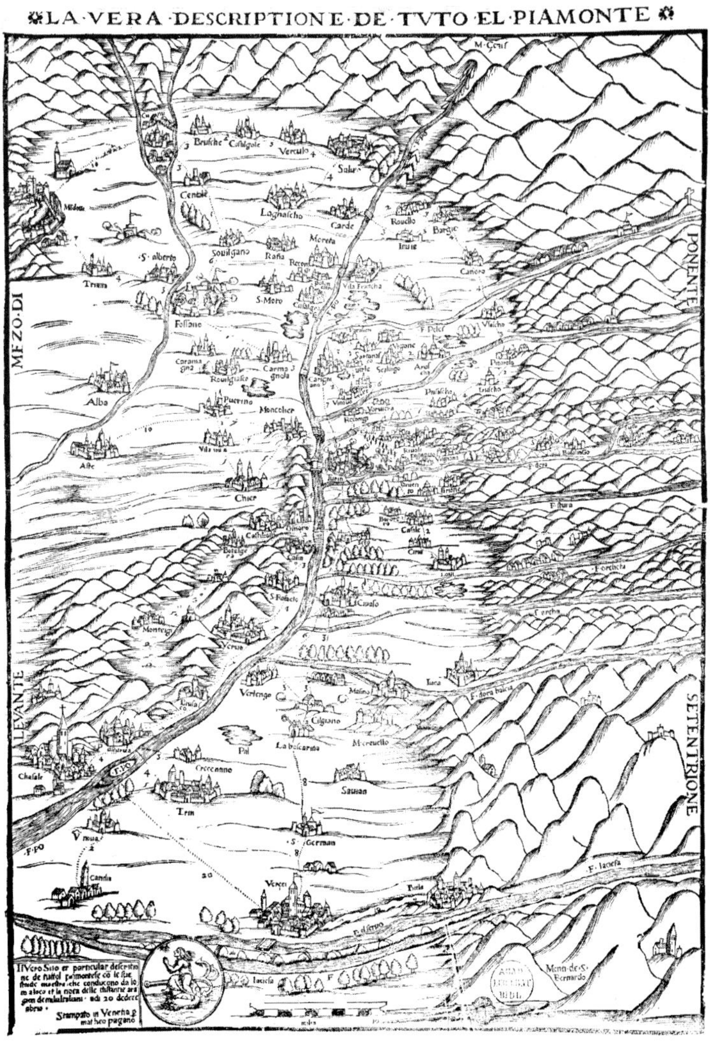

FIG. 13 Matteo Pagano *La vera descriptione de tuto el piamonte* [Venice, Pagano, 1538-39] University Library, Leiden 002-02-009.

1508), as well as the richness of the plates in the Rosselli inventory. In the 1540s, the popularity of the medium was promoted by Gastaldi map of Spain 1544, Sicily 1545, and the Gastaldi edition of Ptolemy's *Geography* 1548, which used an innovative combination of letterpress and intaglio printing including a method of printing the maps' copper plates four at a time.[115] The medium then became routinely used in Italy, not only for single-sheet maps and those destined for composite atlases of Lafréry, Bertelli, and others, but for the remaining sixteenth-century Venetian editions of Ptolemy's *Geography*.

One reason for this gap may lie simply in the poor survival rate for maps—both woodcut and copperplate—when compared to Italian copperplate figurative prints in the period. Very few of the maps listed in Rosselli's inventory can be identified in extant impressions, while Hind lists dozens of Italian surviving intaglio prints from the same period.[116] Another reason that may account for the lack of copperplate maps *vis-à-vis* their woodcut equivalents had to do with the specialisation of the print trade. There was a lack of social association of copper engravers with book publishers during the early sixteenth century, resulting in the separation of the role of the book publisher (using the common press) from that of the printmaker (using the rolling press). Perhaps there were also guild restrictions or city restrictions on who could run which type of press, but no documents have been found to support this. Printmakers shared common interests with artists and goldsmiths, while book publishers, whose scale of production was sometimes very large, were able to establish themselves as independent entrepreneurs and to establish contracts with other trades (for example, blockcutters, papermakers, or binders) as the occasion warranted. Since maps were often planned as parts of books, this encouraged the woodcut medium over copperplate engraving.

★

Although the beginnings of the map trade in the late fifteenth and early sixteenth century were sporadic and studies are plagued by an incomplete record, printed maps in Italy eventually grew popular in the middle of the sixteenth century by becoming available and affordable. As geographical prints, a consumer good, they shared with figurative prints similar technical and trade backgrounds. This modernised their content and freed them from the constraints of classical geography, while the shift from Latin to the vernacular made them more generally accessible. Their markets became less dependent on noble patronage and reproductive methods free from the stylistic and technical constraints of the manuscript and the woodcut. The professional practice of the makers of large-scale maps, the chorographies and topographies, changed radically during the period from that of the painter-miniaturist to the civil-military engineer. At the same time, cosmography became a common scholarly pursuit and an accepted official profession, resulting in the production of improved maps of the world whose structure was still based on the tenets of Ptolemy's *Geography* but whose contents were rooted in the results of firsthand exploration of the world, not on the codified world views of classical and medieval authors.

Although the functions of scholar and engraver had been closely associated earlier in the period (Francesco Rosselli, who frequented humanist circles of Florence and Venice, comes to mind), those functions tended to separate as the trade matured. For example, despite wording his map titles to imply close involvement with the compilation process, Paolo Forlani was probably more a specialised map engraver than an author.

Maps shared with prints the elegant and 'modern' aesthetics of the new technique of intaglio engraving. Given the close association between map and print engravers, it is not surprising that the style of one mirrored the other to some extent. We can discern a 'mapping effect' in which a broad manner and defined style were adopted by printmakers from Rosselli to Sanuto. In this respect, the engraving of lettering made for a crucial difference between maps and prints, as maps required far more annotation, triggering the search for technical shortcuts. In addition, the earlier separation of the crafts of copper engraving and the book trade had left book illustration largely to the blockcutter. Although now among the rarest printed documents of the period, many woodcut maps fulfilled a key role as prototype versions of maps destined for the wider market. There was now a renewed association of the engraving and bookprinting trades that led to a dominance of the intaglio technique in the second half of the sixteenth century.

Printers and Publishers

THE DISSEMINATION of the geographical print in the Italian Renaissance was the métier of the *intagliatore*, *stampatore*, and *libraio* or *mercante di stampe*. Although the success of Rosselli's workshop might have augured a strong map and print trade in Florence for the rest of the sixteenth century, by far the most active centres of printmaking in Italy for the rest of the century were to be found instead in very specific neighbourhoods in Rome and Venice.

ROME AND VENICE

In Rome, the printmakers' quarter was in the Parione (parish of San Lorenzo in Damaso), particularly the Via del Pellegrino and the Campo de' Fiori.[1] (Fig.14) It was a quarter populated by a rich clientele served by bankers, goldsmiths, printsellers, and merchants of devotional objects. The earliest intense period of printmaking in the city was between 1508 and 1527, during which time Marcantonio Raimondi and a group of engravers and printmakers in Raffaello's circle established a reputation for their engravings following the master's paintings.[2] Antonio Salamanca, one of the mainstays of the Roman book and print trade, had already taken up residence in the city in 1517. He was dubbed 'Orbis et Urbis Imitator' (imitator of the world and the

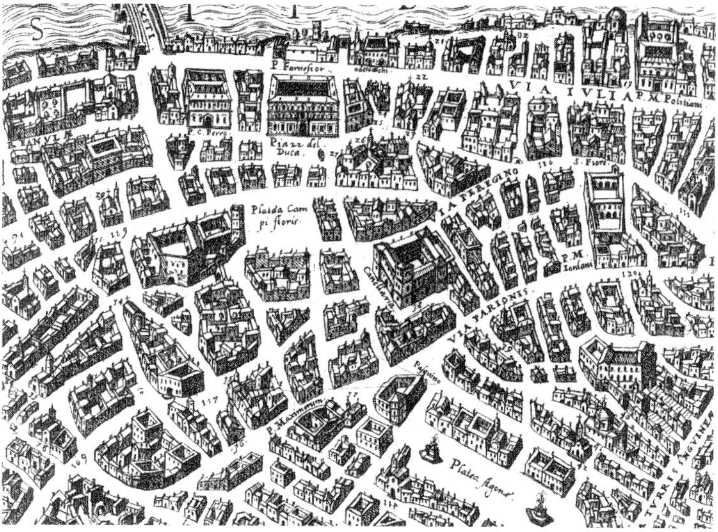

FIG. 14 Detail of the Parione area of Rome on Etienne (Stefano) Dupérac, *Nova Urbis Romae Descriptio* Rome, Antonio Lafreri, 1577. BL Map Library 23805 (8).

city) by his portraitist Nicholas Béatrizet. He may have been a forcibly baptised Jew from Spain, a *converso* taking advantage of ties to both the Christian and Jewish networks around the Mediterranean.[3]

During the sack of Rome in 1527 and 1528, the population of the city was reduced from 55,000 to 32,000.[4] Few businesses survived. Marco Dente, Raimondi's pupil, died in the disaster, possibly defending the shop, and Raimondi himself fled back to Bologna. Salamanca's Spanish origins no doubt helped his chances of survival. The brothers Michele and Francesco Tramezzino fled to Venice, but Francesco returned on 5 July 1528 to set up shop in the via del Pellegrino, which by 1531 had become a meeting place for antiquarians and Roman scholars. The Roman market was eager to absorb maps, plans, and architectural views of ancient and modern Rome as the demand exploded for antiquarian subjects and the results of archaeological excavations.[5]

After the sack, an influx of foreign immigrants enlivened the print trade. Antoine Lafréry arrived from Franche–Comtois in 1544 and set up a partnership with Salamanca that lasted from December 1553 until Salamanca's death in 1562.[6] Lafréry's output increased so significantly during this collaboration, and so intensely in the decade that followed it, that it started a dynasty of map and print publishers in Rome as plates were handed down through generations. (Fig.15) Salamanca's son and heir Francesco, on inheriting his share of the business in 1562, agreed to sell his rights to Lafréry for 1,000 scudi, barely a third of their value, confirming the lack of interest that he had in the business (his name appears on only one engraving, a map of Greece).[7]

Lafréry had the idea of gathering his prints into three bound selections for sale to collectors. The *Illustrium virorum*, compiled by Achille Estaço or Statius, correspondent of Aldus Manutius, and dedicatee of the map of Portugal engraved by Sebastiano di Re and published by Tramezzino in 1566, was a book of portraits of eminent ecclesiastical, political, and military figures. The *Speculum Romanae Magnificentiae* with the frontispiece engraved by Etienne Dupérac, was an album of architectural plans and views of Rome and its monuments and statues. The superbly detailed series of over 600 plates still needs a union catalogue despite several thorough descriptions of individual collections, none of which contains all the examples.[8] The *Geografia*, to which we have already referred, was the geographical equivalent of the other two, containing maps and topographical views.[9]

Lafréry died intestate in July 1577 and his assets were divided into three parts: one third was given to Claude Duchet, Lafréry's nephew, one third to Etienne Duchet (Claude's nephew) and the other disputed third was divided into six parts, most of which included plates locked up until the suit was settled. On Claude Duchet's death in 1585, Giacomo Gherardi carried on the shop in the via del Parione, but on his death ten years later, the tradition of printing with the legacy of Lafréry's plates in that location was broken.[10]

It was a highly competitive business and rivalries were com-

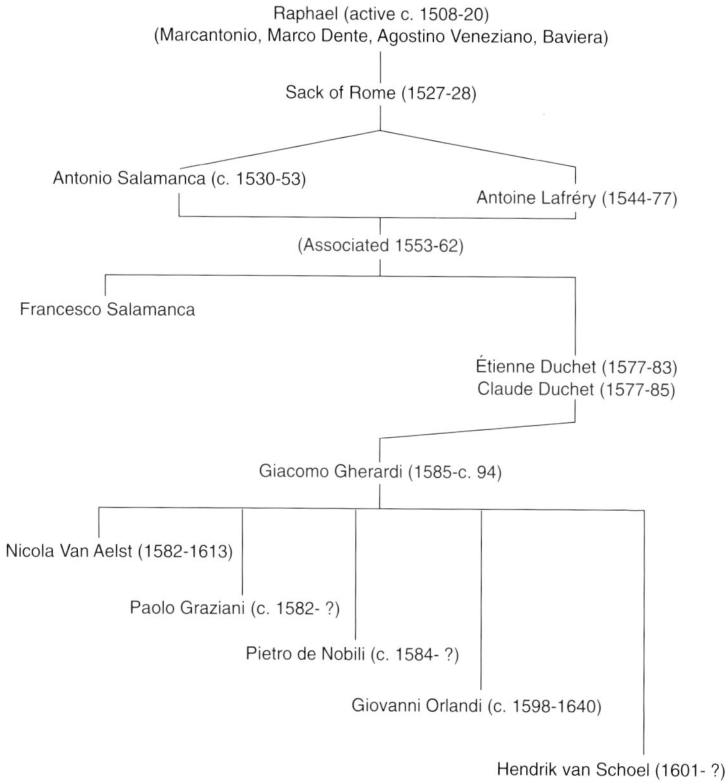

FIG. 15 The lineage of sixteenth-century printsellers in Rome, adapted from Ehrle and Bellini. Diagram by University of Wisconsin Cartographic Laboratory.

mon. When Lafréry died in 1577, Duchet, the rightful senior heir, was in Sicily. The day after he returned to Rome, one of Lafréry's engravers, Gerolamo, disappeared and his body was later recovered from the Tiber. Seven employees from Duchet's shop and that of his rivals Lorenzo de la Vaccherie and Donato Rascicotti, were imprisoned during the enquiry. De la Vaccherie had questioned Claude's right to be an heir of Lafréry's shop, and Claude, for his part, had accused Donato Rascicotti of stealing a large quantity of drawings from their shop. Claude also accused Gilles the Fleming (Gilles Fiamingo), who had been lured away by de la Vaccherie, of stealing work of Cornelio [Cort] worth 70 scudi.[11]

In Venice, the printmakers' quarter was in the Merzaria and Frezzaria, two parallel series of streets that wended their way from the Piazza San Marco northwest to the Rialto.[12] (Fig.16) The Merzaria was lined with retail trades of all types, probably the same number of shops then as today (an incredible number). John Evelyn described it thus:

> Hence I pass'd thro' the Merceria, which is one of the most delicious streetes in the world for the sweetnesse of it, and is all the way on both sides tapistred as it were with cloth of gold, rich damasks and other silks, which the shops expose and hang before their houses from the first floore, and with that variety that for neere half the year spent chiefly in this Citty I hardly remember to have seene the same piece twice exposed.[13]

A critical mass of engravers and publishers who had shops on the Merzaria or neighbouring streets in the late 1550s and 1560s included first Fabio Licinio, then Gian Francesco Camocio, Paolo Forlani, Niccolò Nelli, Domenico Zenoi, Michele Tramezzino, Ferdinando Bertelli, and Bolognino Zaltieri. More map titles were printed in Venice in 1566 than in any other single year in any other city previously. As we have already seen, the blockcutters Vavassore and Pagano had already been active for several decades, but the copperplate engraving trade had only picked up significantly in the 1550s.

The subjects of maps published in 1566 alone show where the demand lay: Of the thirty-six *dated* plates issued that year, eight were of subjects associated with the Turkish wars, nineteen were maps of countries and regions, two were continental maps, and one was a world map.[14]

The relationship between engravers, printers, and publishers in Venice was extremely fluid and informal: plates might be borrowed, rented, held as collateral, or sold; prints could be whole-

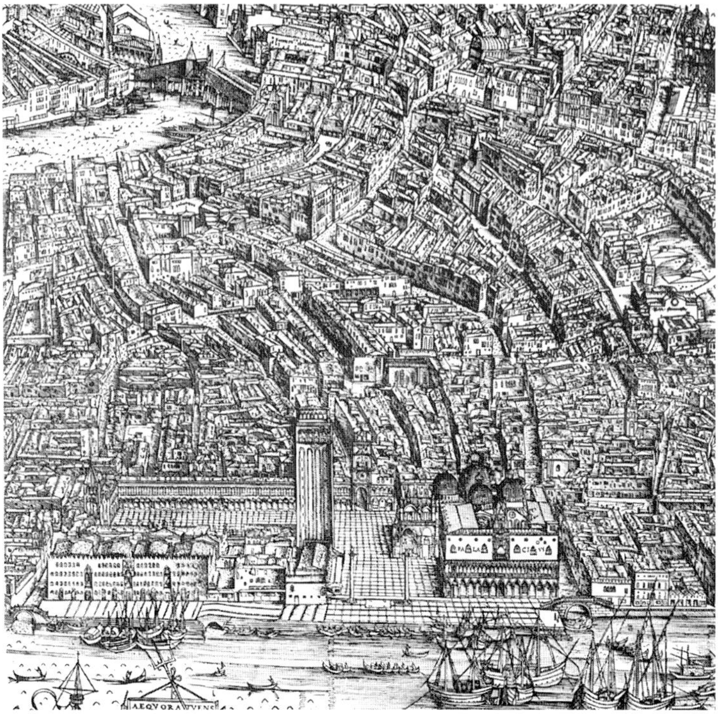

FIG. 16 Detail of the map and printselling areas of the Merzaria and Frezzaria in Venice between San Marco and the Rialto from Jacopo de' Barbari, *Venetia* Venice, Anton Kolb, 1500. BM Prints and Drawings 1895-1-22-1102 . . . 7.

saled from one seller to another and plates were rapidly (and sometimes temporarily) engraved with the name of the distributor. For example, the distributor Camocio is named as publisher on over half of Forlani's maps that name a publisher.[15] In late 1565 or 1566, the prolific book publisher Bolognino Zaltieri purchased some of Forlani's plates (Forlani added the name of the new publisher) including a view of Venice, maps of North America, France, and Ireland, a print of the siege of Tunis and

Goletta, and a print of the Hall of the Grand Council (engraved 8 July 1566).[16]

THE PRINT SHOP

A useful source in reconstructing the procedures and costs of an intaglio printmaker's shop is Vittorio Zonca's *Novo teatro di machine* (1607).[17] The work has been strangely ignored in books on the history of printmaking, and yet it influenced other contemporary manuals in other countries. The earliest manual usually cited is that by Abraham Bosse, published in 1645, a rich source of technical information, but this has now been supplemented by a recently published manual on engraving by Domenico Tempesti from the mid seventeenth century.[18] Zonca's *Teatro* contains the earliest published Italian print of a printmakers workshop, comparable to the often reproduced illustration after Jan van der Straet in 1598.[19] Both prints include scenes of plate preparation, printing, and drying. (Fig.17)

Inking was a time-consuming and tedious bottleneck.[20] Tempesti reports that to obtain really black ink or 'bright French ink', the vine-black had to be specially boiled with beer or white wine (six tankards of beer were particularly recommended).[21] Stiffer grades of ink were used in the summer and with a new plate. After 300 impressions, the ink could be made a little softer. If it were too soft, it produced an ugly red or purple oily impression colloquially called 'fla fla', the result of an overload of oil.[22] However, coloured copperplate inks were known in the sixteenth century, if unusual for maps, so the lavender impression of Forlani's map of Savoy now in the University Library, Leiden, may be an example of coloured ink rather than 'fla fla.'

Before taking ink, the plate first needed to be cleaned and heated. In Zonca's illustration, a shop-boy or apprentice (*garzone*) of relaxed demeanour is indifferently heating a copper plate containing a portrait over a fire of chestnut and hornbeam charcoal. Tempesti reports that one can print 80 or 100 impressions a day, and 50 for a 'large print'.[23] Assuming a single press in the shop, a ten-hour day, and a 'large' print being a *reale* print of about 29.5 × 43.4 cm or 1280 cm^2, it is possible to estimate that

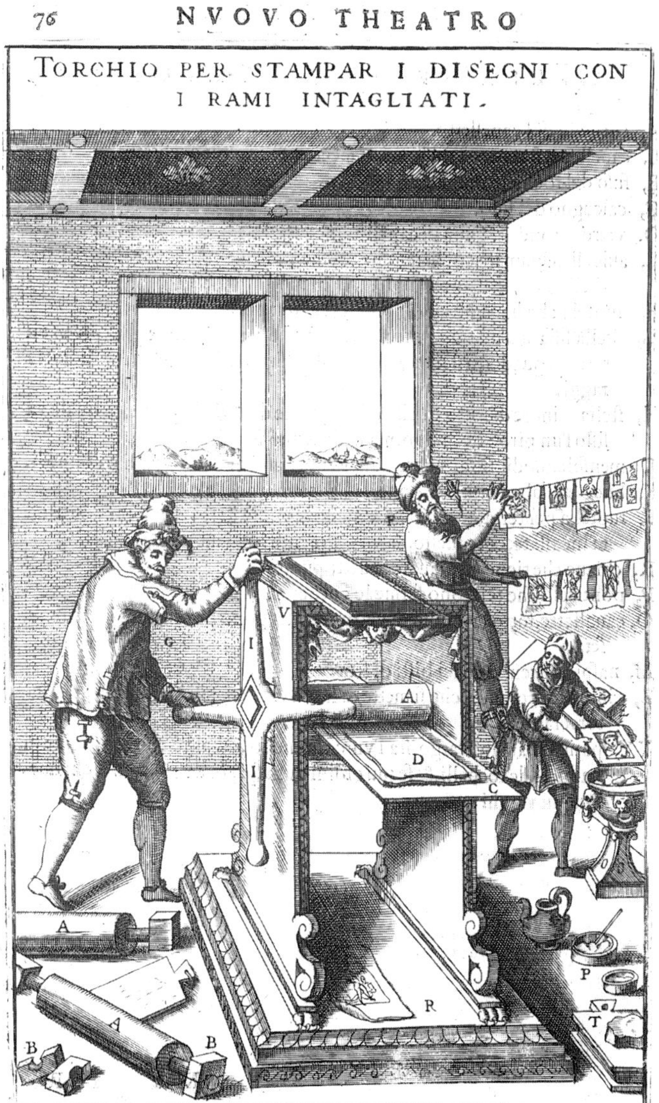

FIG. 17 The printmaker's shop, from Vittorio Zonca, *Novo teatro di machine et edificii*, Padua: P. Bertelli, 1607, p. 76.

it took about 12 minutes to ink such a plate, an impressive rate indeed. Prints with finer detail, such as maps, would clearly have taken longer.

The useful life of a copper plate is a common subject of speculation among map and print historians, necessarily based on modern experience in the absence of contemporary accounts. Estimates of the number of impressions that could be taken before fine lines began to disappear range from 50 to 1500 'usable' impressions.[24] A contract between the architect Juan de Herrera and two Italian printers specifies 4000 impressions from each plate, but there is no corroborating evidence that all these were ever made.[25] A document from the Fontainebleau school of engravers cites 2000 impressions and this is corroborated by Zonca in which it is reported that the life of a copper plate before retouching is 1000 impressions, and after retouching is 2000. For etching, Zonca quotes 500 and 1000 respectively.[26] His figures were repeated in Heinrich Zeising's German version of Zonca's work.[27] For map printing, the tolerance for worn impressions may have been greater than for more artistic prints; certainly more examples of worn copper plates for maps are evident. A systematic study of impressions from a single plate traced through its successive states and correlated with watermark evidence has yet to be undertaken, but such a study might yield important data relating to the longevity and rate of wear of plates.[28]

Paper costs in the sixteenth century are also difficult to estimate, but documents from 1476 and 1622 provide baselines that can give an approximate idea of what printers were paying. Prices in Florence in 1476 for paper of the sort that might have been used for printing included 9 soldi for a quaderno of *reale* size (= £9 per ream), £5 10s for *mezzani bolognesi* size with a ladder watermark, £4 for a ream of piccoli with an eagle or hat watermark, and £3 6s for a ream with the scissors watermark. Five reams and 10 quaderni of *foglio mezzani bolognese* (ladder watermark) cost £30 5s (a figure equal to the £5 10s quoted above), and 6 reams and 18 quaderni of *foglio delle cesoie* (shears or scissors watermark) cost £22 16s (equal to £3 6s per ream quoted above).[29] In 1622 in Verona, the range for paper of *reçuta* size (approx. 45 × 31.5 cm very close to *reale*) was the equivalent of £6 6s to £7 4s Venetian lire.[30]

The paper was dampened to make it softer and more malleable so that the ink could be more easily pulled out of the incised lines in the copper plate. Dampening also increased the life of the copper plate, by reducing friction in the press. The damp paper was held ready for printing between two wooden tablets with handles and weighted down with a rock. These wooden tablets were clearly a characteristic feature of the printmaker's workshop; they are frequently seen on maps as 'ansate' or signature tablets—small nameplates holding text or an engraver's monogram. (Fig. 18)

The copperplate printer was a skilled artisan. Experience was necessary to ink the plate as efficiently as possible, dampen the paper correctly, and then apply just enough pressure in the press, thereby lengthening the life of the plate. Printing was not a purely mechanical job and the wages were commensurate. Figures for payment of intaglio pressmen in Italy in the sixteenth and seventeenth centuries are hard to find, but Consagra has provided several costs per sheet (averaging about 3 to 6 scudi per 1,000 sheets printed) from the shop of the de' Rossi and reports that Giovanni Giacomo hired a pressman for an annual salary of 54 scudi in 1690.[31]

After taking the impressions, an assistant (*famiglio*) (P in Fig. 17) hung the finished prints up to dry. From time to time, they would be taken down and pressed between the boards or in a standing press, a few sheets at a time, to remove the crease caused by the line on which the sheets were hung.[32] If the ink was not quite dry, this would leave an image of the print underneath on the back of the print on top. The bibliographical significance of this unintentional offset is discussed below.

Multiple impressions from the same plate allowed no great increase in efficiency. Since the plate had to be removed anyway from the press after each impression for inking, its replacement by another plate would have consumed little extra effort except for extra cleaning. The printing of copper plates in tandem—one being inked while the other was being printed—was particularly convenient when printing maps of more than one sheet. These

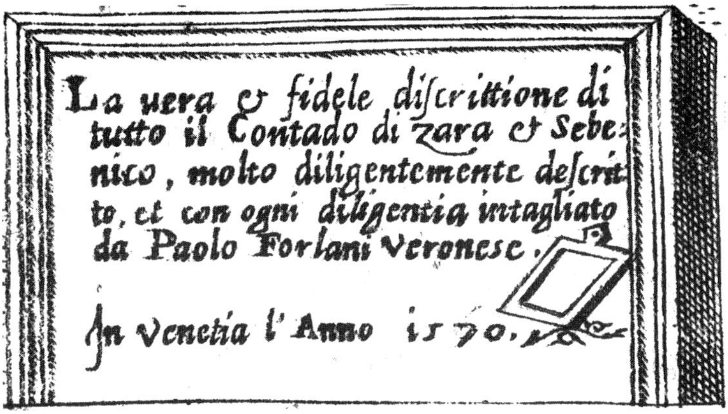

FIG. 18 Ansate tablet on Paolo Forlani, *La vera & fidele discrittione di tutto il Contado di Zara & Sebenico* Venice, 1570. Newberry Library, Chicago, Map 2F205

technical factors account for the fact that 'self-offsets' are rarely found. They also provide evidence for the common practice of printing 'on demand', one or a few impressions at a time as the occasion warranted, rather than printing a whole run of impressions from a plate for stock. In the mid 1560s, Venetian map sellers began to offer miscellaneous bound collections of maps and prints assembled to order; the earliest that survives is the atlas in the Biblioteca Marciana.[33] At least part of the contents of these *atlas factices* would be custom printed from plates in stock.

No doubt small stocks of prints were also kept for sale and trading with other publishers. In the case of the Rosselli inventory (1527), in addition to single impressions of maps (the large majority), there were also multiple impressions: 5 world maps, 6 navigation charts, 19 maps of France, and 20 small world maps.[34] In the 1581 inventory of Claude and Etienne Duchet, 18 impressions of a large 16-sheet map of Europe—no doubt the map by Giovanni Pietro Contarini (1572)—were listed.[35] For the firm of de' Rossi in the seventeenth century, there are many examples of map stocks cited by Consagra including impressions of two-sheet *reale* maps of Naples (24), Genova (100), Venice (45), and Rome (30).[36]

From the figures given, it is possible to estimate the number of impressions of *reale* sheets of a map or print needed to recover costs. The engraving costs for producing the plate in seventeenth-century Rome might include, as a rough estimate, 15 days of rent, taxes, and other overhead of about 6 scudi, 8 scudi for the artist, 20 scudi for the engraver, and 6 scudi for the copper plate, making a total of 40 scudi. For the costs of printing 500 impressions from such a plate, the printer might have been paid 5 scudi, two apprentices 2 scudi each, and a seller 1 scudo. A ream of paper cost about 2 scudi. Printing would have taken about a week, adding another 3 scudi in rent for the presswork. The manufacturing costs thus total only 15 scudi (roughly a third of the creative costs of 40 scudi) for a total of 55 scudi for 500 *reale* sheets of maps or prints.

To weigh this cost against the possible income from selling 500 impressions, we need to estimate the average price range for a map or print of the *reale* size. Many factors apart from the subject, complexity, and production costs influenced the price of a print. Discounts could be given to consumers who bought bound sets of prints, and bulk discounts were available for bundles of prints shipped to other printsellers, as was often done. As a crude estimate, however, we can suggest that a *reale* map or print might be sold for a range of about a quarter to half a scudo each.[37] At this price, the income from 500 impressions would be between 125 and 250 scudi. The break even point in number of impressions sold to recover costs would fall between about 110 and 220 impressions. The minimum number for a sufficient margin of profit would probably have had to be in the range of 250 to 300 impressions.

OFFSETS AND WATERMARKS

The fierce attrition rate of maps and prints in sixteenth-century Italy, particularly those issued as single-leaf woodcuts, many of which survive in only one impression, has already been noted. The lack of technical information in contemporary documents compounds the problem of finding sources to reconstruct the methods used. Any physical evidence therefore that can be marshalled to draw connections between plates or paper used in the production of both maps and prints is essential in reconstructing the trade. Several types of physical evidence can be used to draw out such associations, notably offsets and watermarks.

An offset is an unintentional reversed image set off on the verso of a map or print, usually from being stacked under pressure while the ink was still drying.[38] (Fig.19) Offsets are rarely noted in bibliographies of maps or prints because these documents are matted or mounted in most established collections, and the verso is often difficult or impossible to see. Even where the back of a print is accessible, it is rarely catalogued. For example, an uncatalogued offset detail of a woodcut map appears on the back of Nicolò Boldrini's *Rest on the Flight into Egypt*, a woodcut version of a painting by Titian in the 1530s.[39] (Fig.20) I have been able to identify the map detail as coming from a little-known woodcut map of Verona after Giovanni Caroto, the Veronese architect and painter responsible for the illustrations in a book by Torella Sarayna extolling the virtues of the city. (Fig.21) The map was used as the source for G. Braun and F. Hogenberg's representation of the city in *Civitates Orbis Terrarum* (1586).[40] Caroto's manuscript for the city plan survives in the Verona State Archives.

When analysing the evidence of such offsets, we can use the concept of a 'plate cohort,' to reconstruct, for example, a web of association between Forlani and his fellow publishers in 1566 (our 'typical' year). A 'plate cohort' is a group of copper plates present in the same shop at the same time. (Fig.22) When all identifiable offsets are entered in a table, in which the rows represent the identified map or print and the columns represent the identified offset, any correspondence between map and offset can be noted. Associations between these maps can be ascertained whenever they occur in the same row or column. The largest connected network involved twenty-four individual identifiable maps or offset images. In the plate cohort illustrated in Fig.22, no plate is dated before 1562 or after 1567. Moreover, watermark evidence suggests that none of the pairings occurred

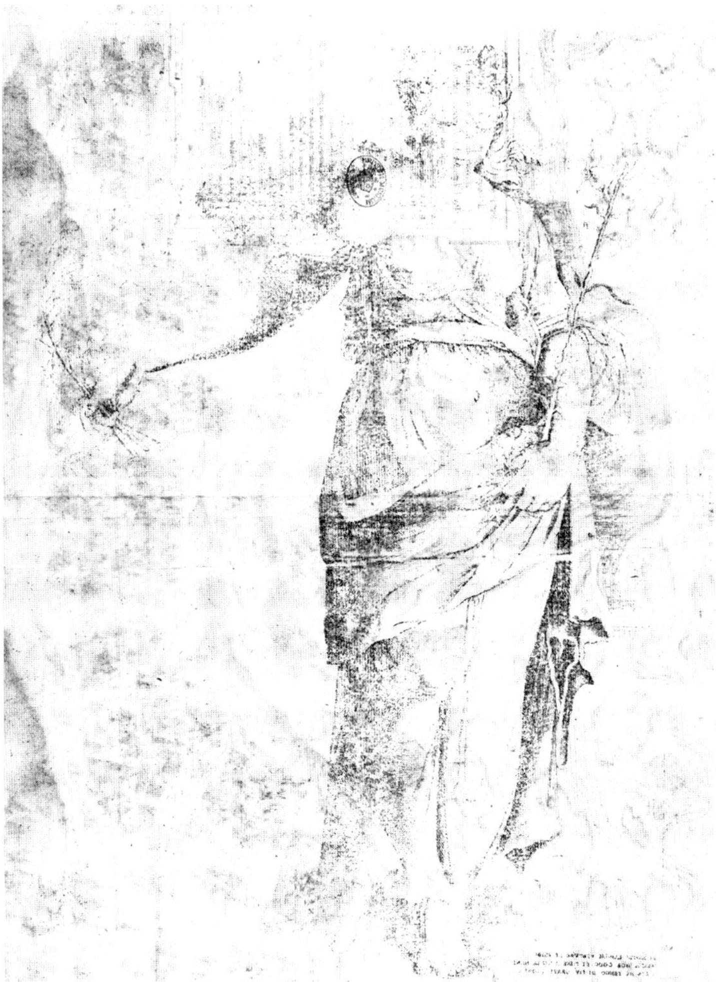

FIG. 19 Offset of unidentified intaglio print of a woman on back of *Ritratto della Fortezza di Ghines*, Rome, 1558, plate 30 in composite atlas BL Map Library Maps C.7.e.2.

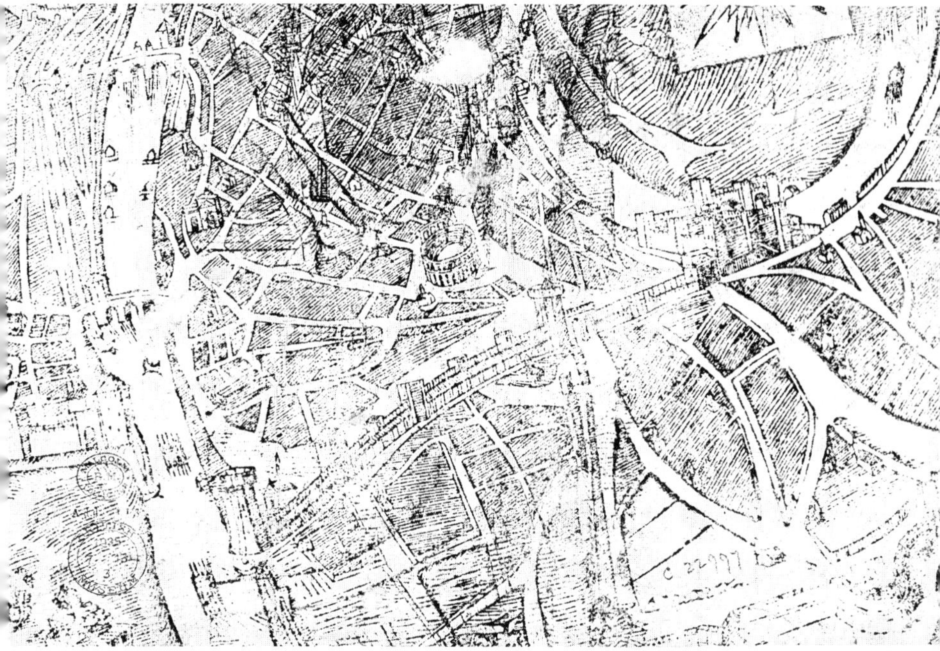

FIG.20 A waste print of a portion of Giovanni Caroto, *Verona* on the back of Nicolò Boldrini, *Rest on the Flight into Egypt*, a woodcut after a painting by Titian in the 1530s. BM P&D: Italian School. Woodcuts. Vol. IV. XV to XVIII Centuries. A–G.

after 1570. In short, most of the pairings are of maps that were printed (according to the date on the plates) within one or two years of each other.

Watermarks are another important diagnostic tool in the study of early maps on paper. The characteristic watermarks of this particular cohort include an oak tree in shield (the arms of the Delle Rovere family and incidentally used by George H. Beans as the imprint for his series of publications in the 'Tall Tree Library'), a double-tailed siren in a circle, and an eight-spoked wheel. All these designs are associated with maps printed

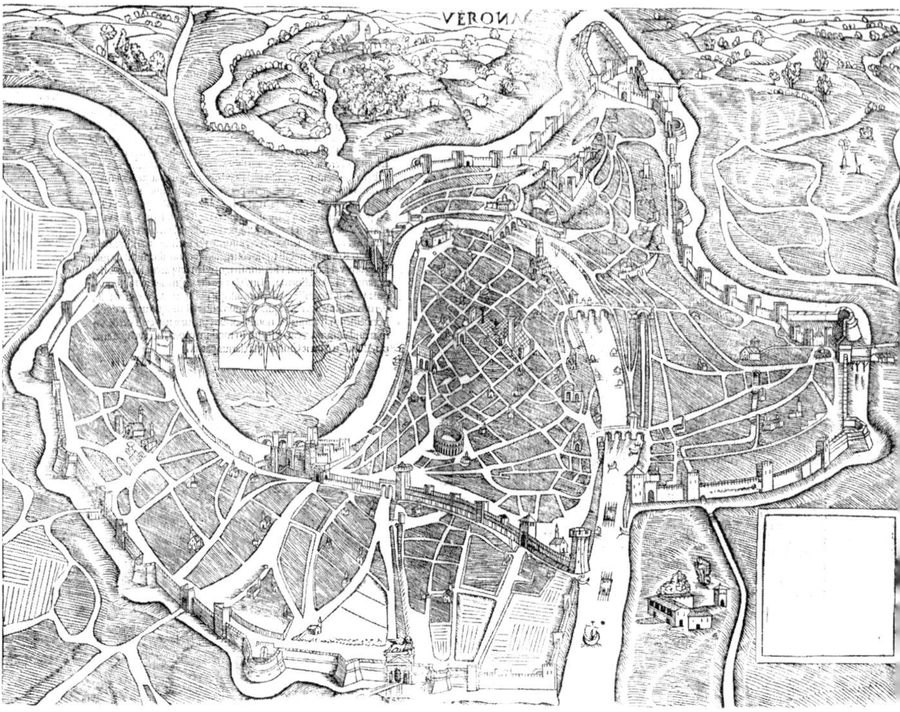

FIG.21 Giovanni Caroto, *Verona* in: Torella Sarayna, *De origine et amplitudine civitatis Veronae* Verona, 1540, pp.39-40 BL Printed Books: 180.k.4.

before 1567. (Fig.23) A particularly strong association can be made between the tree in shield watermark and the map of France published by Zaltieri in 1566, and the second state of Forlani's map of North America also published by Zaltieri in 1566. Out of seven impressions of the map of North America printed on paper with the tree in shield watermark, five have offsets of the map of France. It is sometimes possible to predict correctly the watermark of a map by looking at the offset alone.[41]

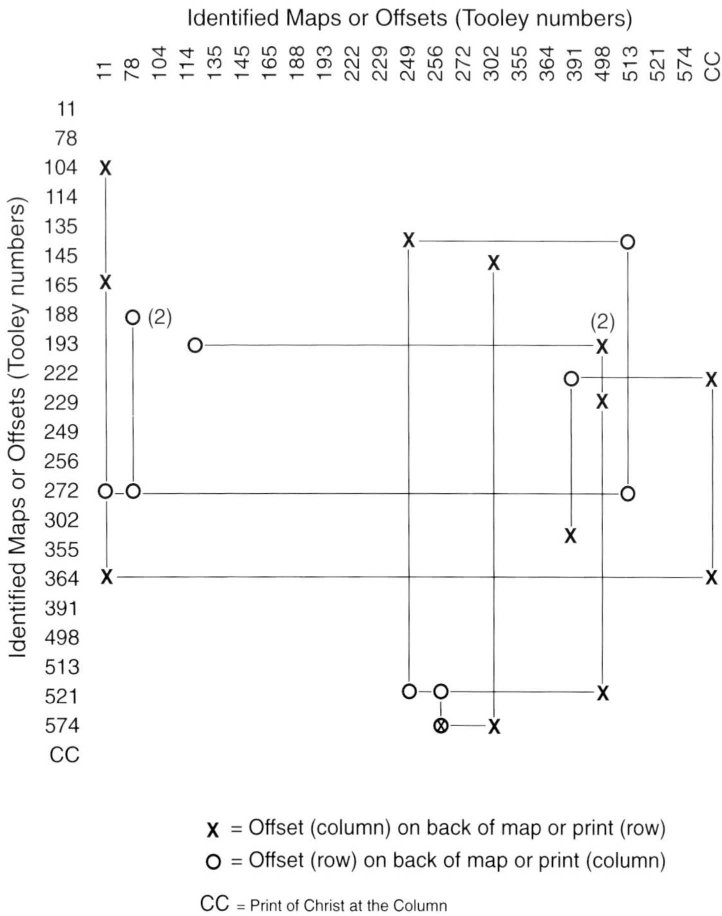

FIG.22 A 'plate cohort' in which maps and their offsets are systematically plotted on a graph to reveal how copper plates have been used in the same shop at approximately the same time. Diagram by University of Wisconsin Cartographic Laboratory.

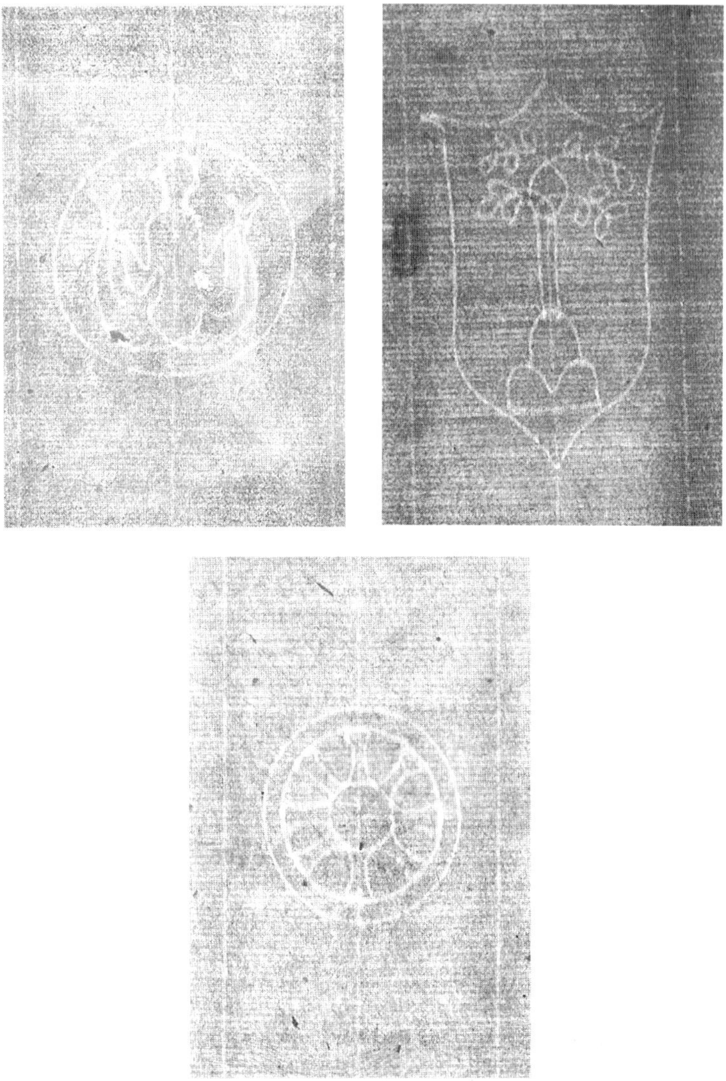

FIG.23 Watermarks associated with the plate cohort in Fig.22, all routinely found on maps printed before 1567. Radiographs by the author.

Waste Prints

When an image on the back of a print is not an offset, it often proves to be a waste print with a quite different image printed on its reverse. One example is a map of the kingdom of Poland printed on the verso of a spoiled proof of the second state of *Diana and Acteon*, identified by Antoine Lafréry's address at the lower right. The printer's possible frustration over this unevenly inked proof is illustrated by his inky palm print planted on it, centre left.[42]

Another waste print so used is one sheet of an as yet unidentified four-sheet Ptolemaic world map in Chatsworth House on whose verso a print of Ugo da Carpi's *The Sacrifice of Abraham* was printed. (Fig.24 and Fig.25) The original block for da Carpi woodcut is thought to have been made about 1515.[43] The *impression* of the map must have been taken before the *impression* of the print, because the latter has picked up ink on one or two of the indentations in the paper caused by the printing of the map. (Fig.26) Such evidence does not prove that the *woodblock* for the map was cut before the *woodblock* for the print, but it does show that the impression of the map block predates the impression of the print in this example. The pairing seems to be unusual, for none of the three examples of the woodcut in the Print Room of the Berlin State Museum has a map or any other print on the verso of any of the sheets.

Waste prints of two Vavassore woodcut maps appear on the backs of two of the eight sheets of Matteo Pagano's woodcut *Procession of the Doge of Venice*, (Venice: M. Pagano, [ca. 1561]).[44] (Fig.27 and Fig.28) Although the identification of this procession with a particular event has not been positively made, the doge resembles Girolamo Priuli (1559-67). The *Procession* was certainly made before 1561, the date of a copy by Domenico de' Franceschi. Printing two of the sheets on waste prints of Vavassore maps suggests possible links between the workshops of Vavassore and Pagano. It may also indicate that maps were regarded as more expendable than prints, perhaps because of their proclivity to become dated.

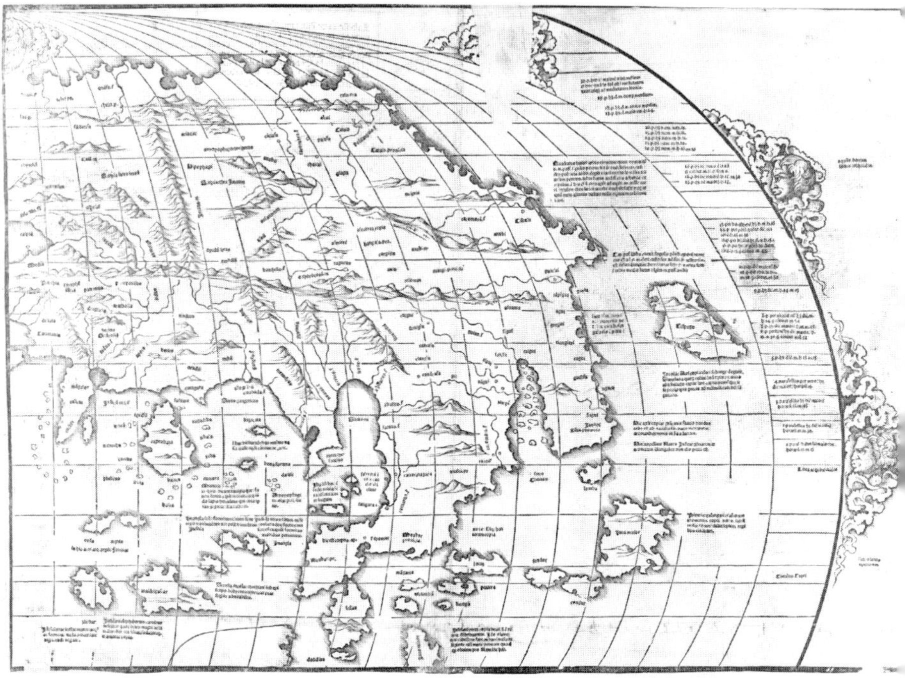

FIG.24 One sheet of an unidentified printed world map on the back of a woodcut of the *Sacrifice of Abraham* after Titian] ([ca. 1520-30]) Chatsworth: Devonshire Collections.

COUNTERPROOF

Quite different from either an offset or a waste print is the counterproof, an intentional wrong-reading image from a freshly pulled impression. Counterproofs were obtained by placing a freshly inked impression of a print against a new sheet of paper and running them through the press together. Read as a mirror image, counterproofs were used mainly as a guide for retouching the copper plate, but they could also be used to reduce engraving time by forming the other half of a symmetrical design, as in

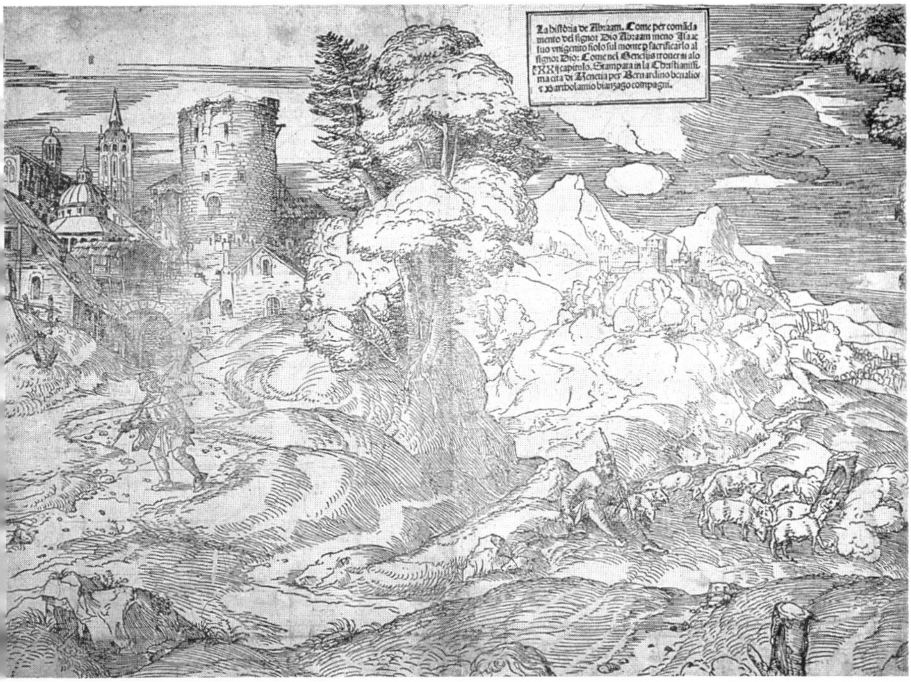

FIG.25 The *Sacrifice of Abraham* after Titian, Venice, Bernardino Benalio, 1515 (a later impression). Chatsworth: Devonshire Collections.

an architectural print of the Colosseum reproduced in Fig.29 where the reversed signature of the publisher betrays the technique.

WATERMARKS

In his paper 'Last Words on Watermarks', Curt Bühler noted that any study purporting to use watermarks to date early documents must assume that the mould for making paper had a limited life, that it was used continuously, that all the paper produced by a mould was used up within two or three years, that the paper was used by the printer in the same order in

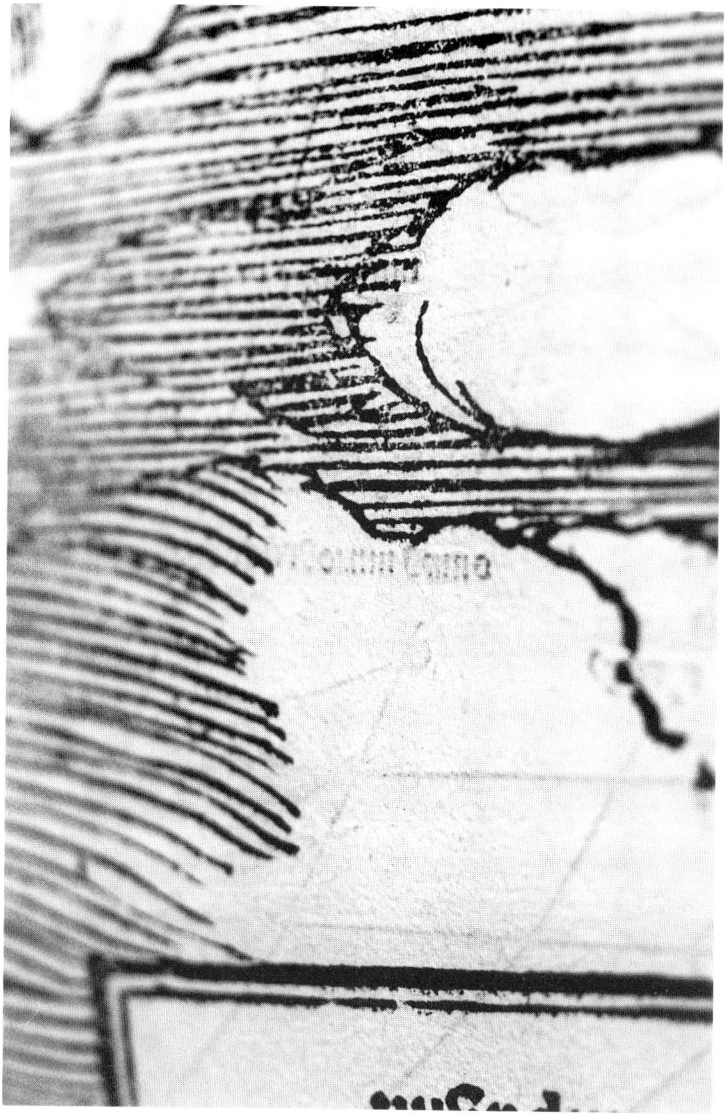

FIG.26 Detail of The *Sacrifice of Abraham* after Titian, Venice, Bernardino Benalio, 1515 (but a later impression), showing ink from the woodblock for the print picked up on the indentations caused by the printing of the map on the reverse. This indicates that the map was printed first. Chatsworth: Devonshire Collections.

which it came from the mould, and that stocks of paper were not accumulated.[45] These objections are difficult to dismiss. Thus, it is difficult to imagine *any* scenario in which a mould was used continuously throughout its life, that paper stocks were *never* accumulated, that sheets were *always* used in the same order out of the ream. Nevertheless, the empirical evidence suggests much more regularity in the use of paper stocks. To judge from the 335 watermark radiographs of unique paper moulds (out of a possible 1,200 examples) from sixteenth-century Italian maps I have studied, it is indeed unusual to find a particular watermark occurring substantially out of the period during which the mould would have been in use.[46]

I have already demonstrated–through a combination of radiography and particle-induced x-ray emission analysis (PIXE)—the high correlation between a watermark and a particular batch of paper.[47] Different papermakers using different recipes of pulp are thus unlikely to have used the same paper mould. In other words, it is unusual indeed for a watermark to appear grossly out of context, as would happen if paper moulds (with their watermarks intact) were freely transferred from one papermaker to another. It also revealed that radiograph matching was a suitable surrogate for the much more expensive PIXE techniques, making the compilation of good albums of watermark radiographs a priority over further PIXE testing.

Caution is still needed, however, as regards the precision with which watermark evidence should be used. Consider, for example, the nearly forty different examples of a key pair of twin watermarks showing a two-tailed siren in a circle under a star. One or another of these watermarks occurs on maps dated 1557, 1559, and every year from 1562 to 1570. Since the earlier maps were printed from plates still available in the late 1560s, it is likely that the life of this pair of watermarks can be narrowed to 1568–1570 or even 1569–1570.[48] Assuming that no map with this watermark dated after 1570 comes to light, and as more examples are found, the likelihood of fixing the end date of the mould at 1570 is increased. Theoretically, it is possible to distribute various states of these watermarks over this two-year period, but such a level of precision is probably unwarranted, as it is unlikely that

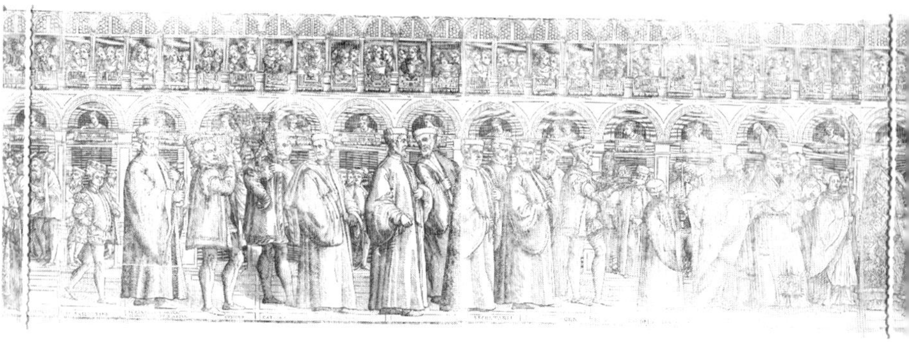

FIG. 27 Matteo Pagano, *Procession of the Doge of Venice*, woodcut (Venice: M. Pagano, [ca. 1561]). BM P&D: Italian Woodcuts. Portfolio 3.

the order of paper use could indicate the order of printing equally precisely.

Composite Atlases

Further contextual evidence for the association between maps lies in the study of watermarks in extant composite atlases of Italian maps.[49] Four atlases are particularly significant, as they are among the earliest Venetian composite atlases, dating from the mid 1560s. The Newberry Library Italian composite atlas, in a full leather contemporary binding with endpapers made of waste paper from a publication printed by Erhard Ratdolt, contains many maps printed from different sized plates on paper of a standard size, 40 × 54 cm (approximately *mezzana*), with eight-spoked wheel watermarks. On the basis of its contents, the atlas has been confidently dated to 1567.[50]

The atlas in the Biblioteca Marucelliana, Florence, in which similarly no map is dated after 1567, bears a siren-in-circle watermark on the marginal strips used to make each map conform to the dimension required for binding. In the Biblioteca Casanatense (Rome) atlas there are likewise no maps dated later than 1567, the majority of its dated maps coming from 1566. The

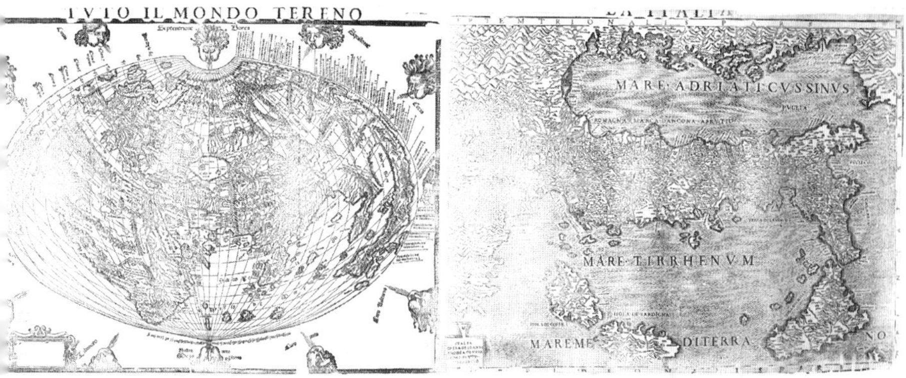

FIG.28 Gioan Andrea Vavassore, *Tutto il Mondo Tereno* [Venice: after ca. 1540], woodcut map printed on the back of a sheet of: Matteo Pagano, *Procession of the Doge of Venice*, woodcut (Venice: M. Pagano, [ca. 1561]). BM P&D: Italian Woodcuts. Portfolio 3.

siren-in-circle watermark is the commonest of several in the atlas and the lack of a siren-in-circle-under-star watermark confirms its pre-1568 date of assembly.[51]

The atlas in the Biblioteca Marciana in Venice is usually regarded as the earliest surviving composite atlas. It contains no map dated after 1565. Lacking a map of the siege of Malta, it probably appeared *before* August 1565. All the blank pages in the atlas contain a watermark composed of a circle with five crescents in a cross under a fleur-de-lys.[52] The dates of binding of these four key atlases provide useful corroborating evidence for the watermark and offset data.

The Licence and Privilege

As with all printed material, map and print trades were under tight control in both Venice and Rome. In Venice, licences were issued by the Council of Ten, the Senate, or some other specially appointed agency, and were intended to control the

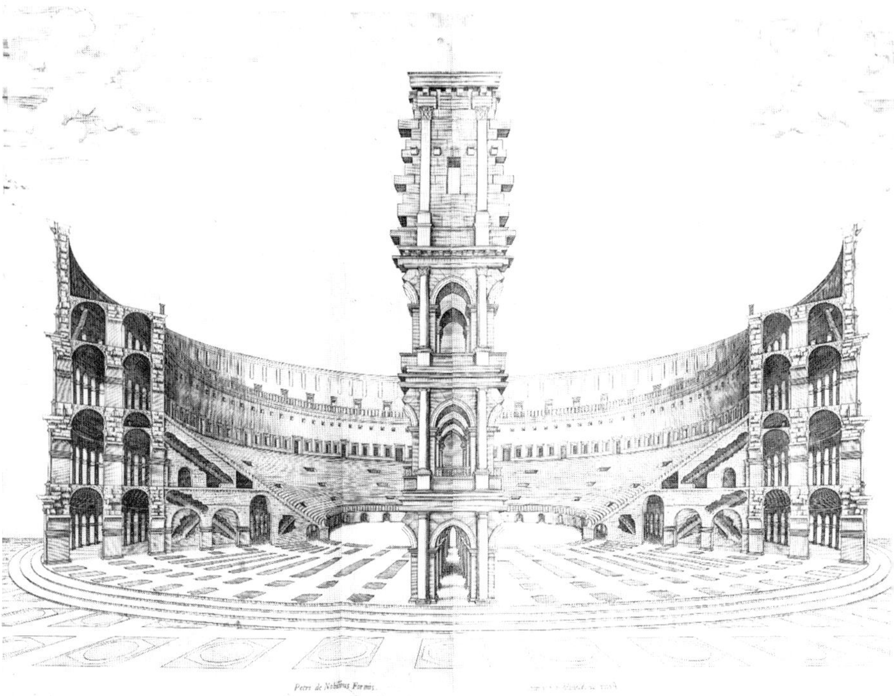

FIG.29 View of the Colosseum, Rome, showing the use of a counterproof in printing symmetrical views. Plate 7 in: Antonio Lafreri, *Speculum Romanae magnificentiae* [1528–1606]. BL Map Library: K.7.TAB.1.

publication of material likely to offend religious, moral, or political sensitivities.[53] Hence, on 7 June 1566, styling himself as a 'libraro and intagliador,' bookseller and engraver of prints and drawings, Paolo Forlani applied for permission to the authorities to put up a shop sign 'at the sign of the column'.[54] 'Al segno della colonna' indeed subsquently appeared on many of his maps. The need for permission reflects the general suspicion that surrounded publishing in Venice after the Council of Trent, as a consequence of which the Venetian authorities insisted, in 1566, that all printed items should be registered without charge by the *Executors against Blasphemy* in a special volume provided for that purpose.[55]

In September 1566, Forlani's rival Domenico Zenoi, like other hardpressed *booksellers* in Venice in the period of the Counter Reformation, sold prohibited books 'sottobanco' and ran foul of the *Esecutori*. The officials duly condemned Zenoi as an 'engraver who has printed some indecent sonnets, and has inserted among them some highly indecent plates, and that he should not leave prison without paying the fine of ten ducats' (perhaps a month's pay of an average artisan—the librarian of San Marco was paid about 17 ducats a month) while Camocio 'libraro at the sign of the pyramid at San Lio was to pay five ducats.'[56] It is probable that these 'indecent plates' were versions of Giulio Romano's pornographic illustrations to Aretino's sixteen sonnets *I modi*, which had attracted the ire of the inquisitor in Rome.[57] There it had been stipulated that no one might sell any printed item without a permit in writing from the Master of the Sacred Palace, the Pope's publications officer.[58]

Licencing was a serious business. In 1542-43 the Council in Venice stipulated penalties for infraction: a fine of 50 ducats for printers and 25 ducats for sellers. Street hawkers of unlicenced material, who presumably could not pay the fines, were to be flogged from St. Mark's Square to the Rialto and imprisoned for six months.[59] The use of a bogus imprint was particularly serious, carrying a penalty of one year in prison and then banishment. In Rome, violators were penalised with the confiscation of books or any other stock and a fine of two hundred scudi.[60]

In contrast to the rigidly enforced licence, the privilege, which was intended to provide publishers with a rudimentary form of copyright, was optional. Not all publishers applied for one, not all privileges were recorded in the *Notatorio*, and records of infringements have not come to light, so they do not contribute to a complete picture of Venetian publishing.[61] In general, however, the privilege records provide the best primary sources and, despite many attempts, we still lack a complete list of relevant licences and privileges.

The privileges do provide valuable information for identifying the date of a map. For example, the privilege for Antonio Floriano's undated world map is dated 18 January 1555 (*modo veneziano*) = A.D. 1556.[62] Some applications may have been made in the handwriting of the cartographer, particularly where the wording of the application specifically includes the word 'io'. Some extant privileges relate to maps now lost or which were never made: 'Gio. Piero de' Marini, *Mappamondo con le nuove scoperte geografiche*, 10 maggio 1529' and another for 'Matteo de' Bardolini *Planisfero*, 30 ottobre 1529.'[63] An exceptional case, which still remains to be properly explained, was the granting of a collective privilege to Domenico Zenoi only a year after his licencing infraction for 'printing or having printed or selling pious figures, portraits, and maps of Europe, Asia, Africa and other separate provinces, such as Italy, France, Spain and others that from time to time will be made, which have not already been made by others in the form and scope that he will make them, for the next fifteen years, under penalty etc.'[64] This action—perhaps an attempt by the Senate to make Zenoi a gate-keeper that they could control—gave him a virtual monopoly of printmaking in the Veneto. However, such a blanket privilege had no discernable effect on reducing the output of other map printers in Venice during the subsequent years.

On the other hand, the terms of a privilege from the Venetian Senate were usually respected: for example, Forlani seems to have waited carefully for the statutory fifteen years to elapse before re-engraving the maps of Poland in 1568 for which Gastaldi had obtained a privilege in 1552.[65] The privilege protecting Gerardus Mercator's 1554 map of Europe appears to have worked, for no copies were made until 1571, and even then the new version was a considerable reduction.

As well as asserting a certain amount of control, official privileges may have been sources of income for both the Venetian Senate and the Vatican. Whatever the fees were, however, they were apparently insufficient to discourage frivolous applications early in the sixteenth century. An application did not mean that the relevant map or print ever appeared. It would seem that authors and printers applied for privileges as a routine to cover the eventuality that someday such a work might be produced. Alternatively, it was abandoned through illness or some other personal circumstances, as appears to have been the case with Paolo da Canal, whose plans for an edition of Ptolemy's *Geography* in 1506 never materialised.[66] Such was the misuse of the system that the Senate decreed on 1 August 1517 that privileges would be granted only for new works and that the right would be lost if nothing was published within a year.[67]

In Rome, the Master of the Sacred Palace not only dispensed licences, but also gave the owners of the plate the sole right to publish the registered print, usually for a period of ten years. As well as the standard copyright privilege, a special privilege could be granted personally by the Pope. Violators were liable to a fine of five hundred scudi, a sum far greater than any imposed by the Sacred Palace.[68]

Dedications

The privilege, as we have seen, was optional. Some important map publishers like Paolo Forlani never applied for one directly.[69] Forlani's lack of requests for 'copyrights' is compensated for by his extensive use of named patrons. This was common practice among publishers, not only to raise capital for special projects, but also to invoke respected names as a deterrent against plagiarism.[70] Forlani's lengthy dedications to prominent individuals often include phrases that stress the effort and cost expended in the production of his maps, implying both that he needed money and that his hard work should not be copied. Another motive was apparently to guard him from possible criticism of his publications, as in his explicit request to Count Geronimo Canossa to defend him against possible critics:

> Coming from such a noble and honored family, I am sure you will undertake to protect the work against detractors, those who don't know how to do anything but reprove the activities of others arising from good and useful intentions . .[71]

The wording of Forlani's dedication to Giovanpietro Contarini (1546–1601) on his map of South America suggests that its purpose was a repayment of some favours, rather than a means of requesting subsidy or obtaining protection against copyists:

> The infinite courtesies afforded me by Your Magnificence cause me to think continually how I might demonstrate a small sign of the infinite obligation that I owe Your Magnificence. So within the past few days I engraved America, with a large part of Florida, commonly called by all the New World, and wished to publish it under the honored name of Your Magnificence.[72]

Tommaso da Ravenna was a fine example of the patron who fitted into the general social web of map publishing. Tommaso was a revered scholar, doctor, and naturalist to whom Forlani dedicated his map of Africa on 9 May 1562. (Fig.5) In 1528 Tommaso went to Venice and a short while later embarked as a physician on a ship in the armada against the Turks. He specialised in 'mal franzoso', or syphilis, and was particularly interested in the medicinal uses of plants gathered around the world.[73] In 1553, he provided the funds to refurbish the church of San Zulian, which was threatened with ruin. In repayment, a statue of Tommaso designed by Sansovino was placed in the lunette above the doorway of the church, showing the master seated on a chair and leaning on the covered table. (Fig.30) He holds in his right hand the *Huysan beata radice* (from the Caribbean) and seems to be explaining its pharmaceutical use to his students.[74]

Tommaso's geographical interests are symbolised in the composition by the representation of a geographical globe, clearly showing the Caribbean from which the prized root comes. Another area of interest to the doctor, Africa, is also shown on the globe as complete. The open book probably represents the medicinal herbal of *Dioscorides*, translated with commentary by Pier Andrea Mattioli of Siena, medical adviser to the Archduke Ferdinand of Tyrol.[75] The statue also contains a suspended celestial globe of larger diameter than the geographical globe and with the celestial equator intersected by the ecliptic and the signs of the zodiac. In such ways, the links between map and patron

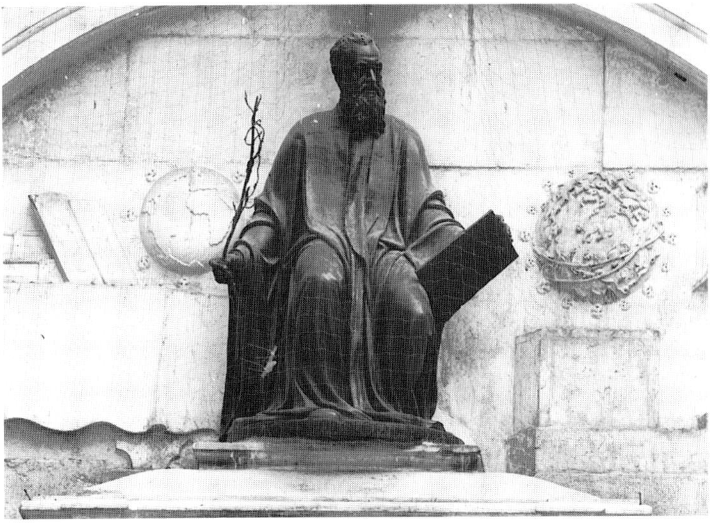

FIG. 30 Statue of Tommaso Rangone by Jacopo Sansovino above the doorway of the church of San Zulian, Venice. By permission of the Soprintendenza per i beni artistici e storici di Venezia.

were extended to many aspects of Venetian academic life: astronomy, cosmography, geography, medicine, botany, and cartography.

★

In the mid-sixteenth century, the map trades of Rome and Venice were more developed than anything Europe had seen earlier in the Renaissance. In Rome, the trade was largely in the hands of foreign immigrants, some of whom established a publishing dynasty that lasted half a century. In Venice, where the organisation of the trade was more fluid, output was larger and the assembling of composite atlases had become an efficient mechanism for marketing maps earlier than in Rome. In perspective, however, the map trade was a small enterprise. Not surprisingly, archival material is elusive compared to that for other trades in sixteenth-century Italy. The study of the print trade in general in sixteenth-century Italy suffers from a similar lack of primary documents when compared to the Low Countries.[76] Even in Northern Europe however, the archival record of the late Renaissance for this facet of the history of publishing can hardly be said to be rich.

Despite the invaluable studies of Antonino Bertolotti, Cardinal Ehrle, and more recently Count Masetti Zannini quoted here, there is still a need for systematic archival searches in Florence, Rome, and Venice. Privileges, and applications for them, remain a potentially useful source for the connections between authors and publishers, even though not all printers applied for one. In the case of Forlani, at least, the dedication seems to have provided at least some protection, as well as furnishing much needed advance capital and endorsing the publication with academic respectability. In Rome, the privileges recorded in the *Brevium Secretorum* (Archivio Segreto Vaticano) remain to be systematically searched. And although a good start was made in Venice by Roberto Almagià and Horatio Brown, the key to finding other documents—contracts, wills, inventories, and other legal documents—still lies in first identifying the lawyers associated with the engravers and publishers and then conducting a long patient search—months at a time—in the relevant files of the State Archives. Until then, our knowledge of the technical processes of map printing and publishing in this period, reliant on serendipidous archival finds, will progress only slowly.

Meanwhile, bibliographers must base their observations on the maps and prints themselves and look for corroborating evidence of associated trades of the goldsmith, papermaker, and book publisher. The significant overlap in the lives of engravers, printers, and publishers of maps and other prints leaves no doubt that the printed map was regarded as a geographical print, subject to similar technical constraints and marketing considerations. At the same time, maps may have been viewed as less prestigious and expendable than their more figurative counterparts. Where maps and prints are found together, printed on opposite sides of the paper, the map was always printed first and then used as a waste print.

There is much value in the use of physical analysis to illuminate connections between maps and prints that otherwise appear to be unidentifiable and unconnected. We need to be sensitive to the physical form as well as the intellectual or stylistic content of documents, literally holding them up to the light and scrutinising their obverse sides for scraps of evidence. The use of offsets and watermarks as bibliographical evidence is still new to the staffs of many libraries; it is almost impossible to request a photograph of the *back* of a map or print. Curators of both map and print collections should be aware that the back of the object may bear bibliographical evidence. A systematic unmounting project would be too daunting, but a start could be made by photographing bibliographical information on the back of newly acquired maps and prints and by registering their watermarks.

Despite widespread scepticism about the value of watermarks in bibliographical research, my own studies have shown that a systematic collection of good radiographs or other objective images of watermarks from separate maps and prints can help reveal the links between map publishers in the Italian Renaissance, for which watermark images are plentiful. Radiography—time-consuming, cumbersome while travelling, and possibly marginally hazardous—may not be the best technique to acquire and analyse these images. With the explosion of digital imaging methods, a rapid, safe, clear, and versatile method of gathering the images must surely soon appear.

Finally, instead of viewing map engravers and printsellers as simply names to be listed in a bibliography, we should place their daily experience into the relevant social context and draw out contacts with colleagues (sometimes fierce rivals) and consumers. For, without a market, these maps, atlases, prints, and print albums would not have been made. Nor would a market have existed without the development of a taste for geographical prints. Our next task is to identify who bought these geographical prints, and why such a taste developed.

III

Consumers

It is curious that so little attention has been paid to both the economic and aesthetic issues involved in the historical consumption of maps as geographical prints. Part of this neglect derives from the biases of traditional economic history toward the supply side and away from the demand side: there have been many books on map production but few on their consumption, thus giving only half the picture.[1] One reason for this is the view of some economic historians that the widespread formation of consumer tastes for material goods did not have roots earlier than the eighteenth century, at which time a sudden revolution occurred. Others disagree, pointing to clear evidence of already well-formed canons of taste for the decorative arts in the sixteenth century. Thus Goldthwaite points out that:

> The success of the decorative arts sector was above all in the way it transformed the social structure of the economy. The channeling of wealth into luxury consumption resulted in its transformation into human capital—an increase in the number of skilled artisans, a rise in the level of their skills, a seemingly infinite expansion of the range of their skills, and a constant stimulus to the imagination with which they executed those skills. The immediate economic effects were the rise of per capita income among artisans and an increase in consumption brought about by the larger number of them.[2]

Goldthwaite goes on to stress the centrality of the decorative arts in this development, using Florence as an example:

> Little is known about the minor arts in Florence, despite an extraordinary abundance of documentation and numerous surviving examples of that production too often stacked away in museum storehouses. As an index to the history of taste they deserve more attention than they have yet received from students of the so-called 'finer' arts. For the economic historian the history of taste is an important part of the history of demand, and until it is written we cannot have a full understanding of the development of one of the most vigorous sectors of the economy of Renaissance Italy.[3]

As an integral, possibly central, part of this activity, the market for prints in Renaissance Italy was also strengthened by the collecting mentality of the sixteenth century.[4] Vasari, whose writings codified the canons of taste for collectible art, rated prints lower as the trend toward popularisation and reproductive engraving increased during the sixteenth century.[5]

In the history of cartography, more attention has been paid to the process of compiling and using maps in official administration, exploration, and warfare than to their purchase and use at the popular level as informal consumer goods. There have been attempts to explain why certain periods of history or certain cultural groups have been 'mapless' or when maps became normally used as political and social tools.[6] Clearly, the content of geographical prints had their roots in the manuscript maps that were made for administrative and scholarly uses, whether in the traditions of the painter-miniaturist, the military and civil engineer, or the humanist scholar. Since geographical prints were based less on direct observation of the world than their manuscript counterparts, it is more difficult to claim for geographical prints a direct pragmatic influence on decision making. Their functions were more nebulous. They helped to form in the minds of a wider audience an image of the geographical context of events in a rapidly changing and expanding world. Equally important, maps as objects displayed in a domestic setting became icons projecting the owner's worldly status. Likewise, scholars and professionals added albums of printed maps and figurative prints to the burgeoning libraries in their *studioli* for private study and academic discussion.

Chandra Mukerji has argued that the study of pictorial prints in the fifteenth and sixteenth centuries is one of the most effective ways of analysing the historical development of consumerism. She poses a theory that consumerism helped to create the capitalism it is conventionally supposed to have followed. She argues that we have been understandably too distracted by the richness of Renaissance painting, sculpture, and architecture to consider the possible effect on the culture of capitalism of the 'humblest and most innocuous items that our culture has produced: pins and laces, pictures and cooking pots.'[7] Some historians have discounted the impact of figurative prints or printed maps on the spread of knowledge of the new discoveries in the fifteenth and sixteenth centuries:

> Lucien Febvre and Henri–Jean Martin argue that the European reading public's image of the world was by and large unaffected by the new geographical discoveries yielded by European expansion. They contend that the public must not have been interested in this material because many surviving manuscripts (both charts and writings) were not put into print. In making such an argument, these authors are assuming that whether works did or did not appear in print in the sixteenth century depended on popular interest (a dubious assumption even in the present day) and that only massive dissemination of information could create an effect.[8]

Mukerji goes on to point out that there was indeed an informative role played by these documents. More importantly, however, these writers have ignored the role of maps and prints as economic objects to be traded in the new consumerism:

> Febvre and Martin reach such a conclusion because they assume that printed geographical works could only have cultural or intellectual effects on European life. If they had considered the economic uses of these documents, they would have realised that, precisely because these items could be used as capital goods,

they did not need to reach a large audience to stimulate social change. They only needed to catch the attention of those merchants or aristocrats interested in sponsoring new forays into the world's oceans. By mass-producing maps and chronicles, publishers increased the scale of the materialism associated with overseas trade, and by trading extensively in these goods, they brought many of the new documents from early voyages to the eyes of interested audiences, breaking the Iberian monopoly over the information they contained.[9]

In the sixteenth century, or in any century, it was not usually the most intrinsically 'correct' or informative geographical representations that reached the public. W. F. Ganong cautioned—perhaps too stridently—that sixteenth-century Italian printed maps 'have no historical or geographical worth', but he acknowledged that 'they do have their interest in the history of cartography, as illustrating some of the methods of its evolution, especially the formalism that intrudes when further data fail.'[10]

What then did maps as geographical prints mean to the society of the sixteenth and seventeenth centuries? As they became more and more a part of everyday life, they played a subtle but important role in the shaping of ideas about the world. Beyond conveying knowledge—factual or otherwise—about strange places and events, they symbolised through a complex iconography some overarching themes: the magic of capturing the world as a single ordered image, the replacement of the content of classical geography with a 'modern' geography that incorporated 'the new discoveries,' and the secularisation of the world image from the representation of spiritual to geometric space.

The precision with which printed maps formed the public's image of the new discoveries is difficult to document. Since early Spanish and Portuguese exploration relied only on manuscript charts, these were either produced before widespread map printing developed or were confidential documents for which printing was out of the question. Even Diogo Homem's chart of the Mediterranean—hardly an avant-garde geographical document—was not printed until 1569.[11] The working charts of trading companies and military establishments were to remain in manuscript much longer than that. It is true, as Febvre and

Martin have said, that the most up-to-date geographical information was not available through the printshops of Venice and Rome, and it is equally likely that the consumers were not particularly discerning as to what was really up-to-date. It is hard to imagine, however, given the burgeoning interest in purchasing topographical and geographical images at the time, whether for display, study, or topical information, that the wide dissemination of the *idea* of the map during the second half of the sixteenth century had no effect on the way the middle classes viewed their world.

DISPLAY

Household inventories provide a useful source of evidence about the market for printed maps.[12] Unlike the Netherlands, where inventories have been used extensively to show the domestic use of maps in the seventeenth century, similar information on the Italian ownership of maps and prints has been all but ignored. This omission is not surprising when we realise that studies of the patronage and ownership of even the 'fine arts' of painting and sculpture have lagged behind biographical and aesthetic considerations in art history.[13] However, for fifteenth- and sixteenth-century Florence, Lydecker has demonstrated two valuable archival sources: the estate book (*libro d'eredità*), which was a specialised form of record kept by an executor to record the disposition of the deceased's earthly affairs, and—perhaps more useful—the collected inventories among the records of the Office of Wards (Ufficio dei pupilli).[14] Scattered among these and other inventories are surprisingly numerous references to maps and prints, often indicating the location in the house where they were displayed. In Venice, there is a collection of registers of sixty-six lawyers recording inventories of household goods of all social classes in all neighbourhoods of the city, compiled during the last half of the sixteenth century.[15] These inventories were usually prompted by the death of the owner, but could also be undertaken to value household goods on the occasion of a new marriage for the support of children from the previous marriage.[16] It is unfortunate that similar inventories are not available for the first half of the sixteenth century to document the display of maps and prints in houses of the lower classes at an earlier date.

Maps of the world, or globes (the word *mappamondo* can mean both) were evidently frequently displayed. In Venice, Marin Sanudo mentioned that a visitor came by on 5 December 1511 to see his studio and the 'mapamondo'. An inventory dated 12 February 1557 for the property of Ugolino di Luigi Martelli notes a multi-coloured world map on canvas in a fir frame; a map of the world in the shape of a heart (very probably Cimerlino's world map) (Fig.31); and in the study above this room, a pair of terrestrial and celestial globes with fittings and a small brass sphere. In 1567, Gabriele Vendramin owned 'un mappamondo turchin in balla.'[17] Perhaps this was the so-called 'Hadji Ahmad' map in Turkish on a heart shaped projection published in Venice by Marc'Antonio Giustinian in 1560. No impressions are known from the sixteenth-century blocks, which were rediscovered in 1795 and are still preserved in the Museo Correr, Venice. If our identification is correct, this entry might provide evidence that the map was not entirely suppressed until the eighteenth century as has commonly been reported.[18]

Federica Ambrosini has documented numerous houses containing continental map sets of Europe, Africa, Asia, and America beginning in the last quarter of the sixteenth century and throughout the seventeenth. The frequency of these references to the 'description of the world in four parts' might suggest that they were a cosmographical accessory projecting the owner's status as a knowledgeable and worldwise citizen. And although such accoutrements were usually the privilege of the well-to-do, the working classes were occasionally able to afford the trappings of the rich. Andrea Bareta, who died in 1587 and who was apparently a wool worker, owned a small collection of pictures including the four continents: 'Asia, Africa, et Europa, et Peru' among the more expected sacred themes.[19]

The 1576 inventory of Gasparo Segizzi, an illuminator who lived in a large eight-room house in the parish of San Barnabà, permits a reconstruction of a complete plan of his house. In the lower portico, there were nine pictures, among them a 'portrait

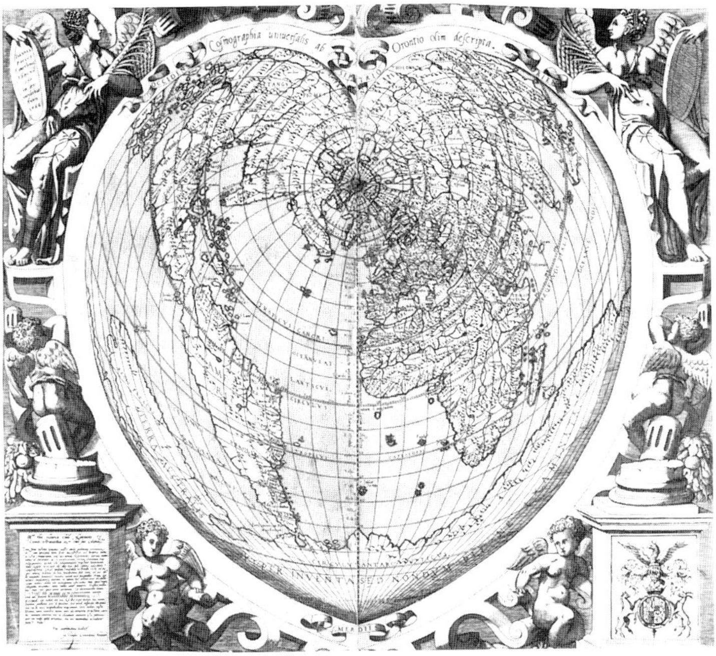

FIG.31 Giovanni Paolo Cimerlino (after Oronce Finé), *Cosmographia universalis ab Orontio olim descripta*, Venice, 1566. BL Map Library C.7.e.2.

of an ancient emperor' and some nudes. In the living room there were fifteen pictures. The long list of pictures is closed with a 'napamondo vecchio' (an old world map).[20] Whether or not 'old' here means in bad condition, out of date, or a map of the Ancient World, we do not know. Twenty-four pictures clearly constituted a substantial collection given the relatively lowly station of their owner.

A collector's collector was Girolamo Gualdo, whose description of his collection survives in an as yet unpublished manuscript in the Biblioteca Marciana.[21] In one of the rooms was a terrestrial and celestial sphere and two other similar ones in an adjacent room. A brass sphere hung from the ceiling in the study in which there were various other celestial and terrestrial globes and a *planisferio* (presumably an armillary sphere). The study ceiling was frescoed with a cosmographical theme.

Ambrosini argues that the clergy were often indifferent to displaying maps and pictures of exotic places, with some rare exceptions. The bishop Leonardo Mocenigo (died 1623) had 'un quadro de napamondo' among the paintings and statues in his Ceneda palace. In 1641 'le quattro parti del mondo in quadri sopra tella et carta' decorated 'un studietto' in the house of Giovanni Rossi, vicar and pievano of San Bartolomeo. Another vicar of San Bartolomeo, Giovan Francesco Montanari, counted among his goods (1697) 'tre parti del mondo in carta' in a room of his house in Fiesso, a fourth part of the world in another room, while the four parts of the world in paper were the motif in another 'saletta'. However, these three privileged ecclesiastics cannot represent the taste of their less wealthy colleagues.[22]

More often, maps were owned by the secular professional classes and nobility. According to a 1602 inventory, the lawyer Ludovico Usper exhibited 'uno napamondo in quadro grande' in the canal entrance of his house at San Bento and owned two terrestrial globes and a celestial sphere. The collector Federico Contarini had 'two world maps' in the 'portego' of his house in 1613 and possessed a map engraved in ivory on a box among other curiosities.[23]

Another way of demonstrating one's cosmopolitanism was to have a collection of smaller maps representing single countries or city views, usually capital cities such as Rome or Paris or metropolises that had achieved fame in the commercial world: London, Amsterdam, or Antwerp. In the Rezzonico palazzo, lined up along the walls of a study and in the corridor next to the dining recess, there were eleven geographical maps representing the parts of the world. According to the inventory of his goods gathered on 30 November 1655, the doge Francesco Molin had collected thirteen 'small sheets of drawings of the parts of the world' and 'various large and medium size pictures with maps of the countries of the world,' apparently including views of regions other than Italy or Europe.[24] Innumerable other cases include selected maps of countries, regions, and cities.[25]

The inventories reveal a marked absence of local views of Venice or icons of Saint Mark, a strange *lacuna* considering the obvious pride with which the view of the city by Jacopo de' Barbari must have been displayed in the administrative offices of the city in 1500.[26] Perhaps the Venetian, so accustomed to experiencing the wonders of the city in reality, felt that a representation of it was superfluous. Why have a map when you have the territory? In any event, the large numbers of images of the city created by the map and print trade of the mid to late sixteenth century were apparently far more likely to be sold to visitors than to the local population.

Although one can find people of the poorer classes displaying a modest map next to the ubiquitous sacred images, the geographical prints in the inventories occur mostly in the houses of the gentry. These appeared not only in Venetian townhouses, but also in the country houses of the Terraferma. The frequency was much greater for the seventeenth than the sixteenth century. Lorenzo Tarabotto kept a set of four continental maps in black frames in one room and a large map of the world in a black pearwood frame in another in his house at Sambruson near Dolo in 1661. In 1663, the inventory of the goods of Giovanni Gottardo Catti at Lancenigo included eight maps of various parts of the world. Large planispheres occupied the entrance rooms of the palazzi of the family Gradenigo at Villarazzo and of Girolamo Borella at Valdobbiadene in 1667 and 1684. The Tasca family exhibited a particular passion for maps in its villa at Gardigiano:

> in the *portico del sopra*: four maps showing the four parts of the world
> in the *primo mezado*: two maps of the four parts of the world
> in the *mezado* near the steward's kitchen: four large maps of the four parts of the world on paper glued to linen
> in another location: four small globes and a large one on its pedestal.[27]

For Rome, Consagra has noted the growing demand in the seventeenth century for topographical images to decorate houses, often the *casini* or country houses on the outskirts:

> In 1665, Alexander VII commissioned Giovanni Giacomo de Rossi to provide printed maps and views for a gallery in the papal apartments of his summer residence at the Castel Gandolfo. It was to be decorated with no fewer than 176 printed maps depicting Europe, the Americas, as well as views of Rome, all hand coloured and glued on canvas by various printmakers and publishers.[28]

A common location for these geographical prints was the entrance hall of a private house, where they suggested that the owner was a person of taste and familiar with the world of scholarship. Bartolomeo Falier kept his maps of the four parts of the world next to his Turkish jewels there. In 1605 the silk merchant Vito Vidotto in San Simeon kept his in the portico with his silver and a spinet. In the latter half of the seventeenth century they increasingly became distributed around the house—on landings, stairways, living rooms, and general purpose rooms.[29]

Lest it be thought that the typical Venetian, Florentine and Roman villa was overflowing with maps, it is worth emphasising that their frequency was far less than that of religious or classical themes. The Venetian owners, for example, were more inclined to display religious prints to confirm the piety of their families or portraits of forebears to demonstrate their noble roots. Portraits of foreign kings or illustrious historical personages, landscapes and marinescapes—often Flemish—were also favourite subjects.[30]

It is worth stressing the Flemish connection, since by the seventeenth century the Low Countries had captured a much larger share of the map market than the Venetian and Roman publishers. The prestige of the map as an artifact to be displayed was clearly enhanced by being a valued import. Although the links between the Low Countries and Italy in the late sixteenth and seventeenth centuries have been briefly alluded to in the literature, much more archival work needs to be done to trace the channels of contact.[31]

In general, the inventories reveal that it was still the wealthy who bought maps in sixteenth- and seventeenth-century Italy, whether they were merchants, art lovers, scholars, or ecclesiastics. The emphasis on the frames—colour, type of wood—clearly indicates that the people taking the inventory regarded the framed maps as furniture. In the case of Venice, it remains to be explained why those who had the means and the intention of investing money in pictures chose cosmography and other geographical themes, which was unusual considering the traditional Venetian tastes of the seventeenth century. The devotional print, projecting the piety of its owner, had a larger and more secure market, but the geographical print displayed on the wall also projected a number of attractive images for its owner. It reflected cosmopolitanism and cultural refinement; patriotism to city, region, or country; humanistic links with well-known places and peoples of another time, and the pure aesthetic pleasure of escapism, both geographical and historical—particularly of evoking past glories of the Venetian republic or the founding of Rome. The geographical print showed its owner to be *au fait* with the issues of geographical discovery and world trade. The presence of maps of the world and the continents hanging on the walls also served to remind Venetians of the republic's former power and perhaps even to trigger dreams that the power might be regained.

The clear integration of geographical and cosmographical themes into the print repertoire by the early seventeenth century is illustrated by the iconography of a print found in a composite atlas once belonging to Isaac Vossius and now in the library of Leiden University. (Fig.32) The print is in the genre of a thesis print, or *conclusione*, which graduating students used as a kind of diploma, replete with punning allegories to display their humanistic learning of classical iconography.[32] A letterpress list of statements to be defended in their thesis would have been attached, but these are sometimes missing from the extant impressions. The prints are often found in multiple states recording the addition of custom-made details required by the student, such as the arms of their patron or *collegio* or some other allusion to their field of study.

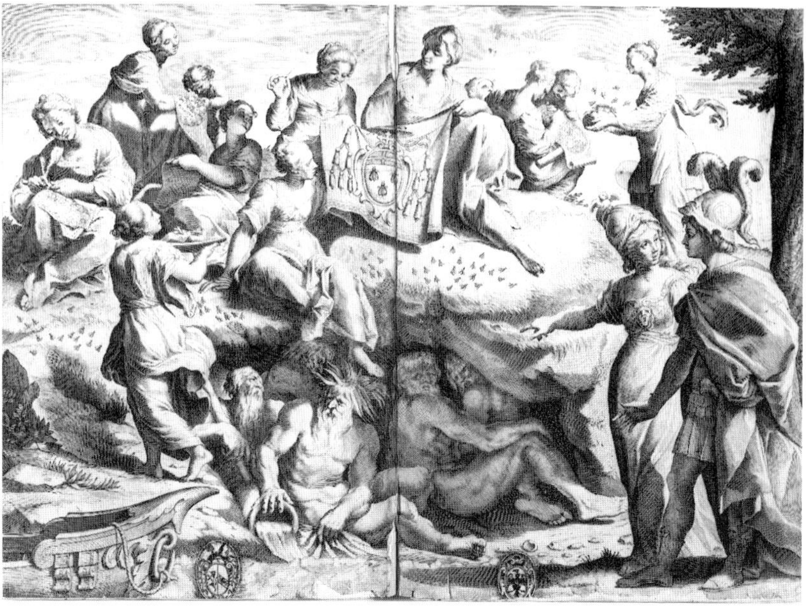

FIG. 32 Johann Friedrich Greuter (after Giovanni Lanfranco), thesis print or *conclusione* made for an unidentified student, dedicated to Cardinal Francesco Barberini, ca. 1625. University Library, Leiden, Isaac Vossius collection, atlas 6, plate 58.

The print is by Johann Friedrich Greuter, son of Matteo Greuter, after the artist Giovanni Lanfranco and probably dates from the 1620s.[33] The main arms are those of Cardinal Francesco Barberini (nephew of Matteo Barberini, who was elected Pope Urbano VIII in 1623, and brother of Cardinal Antonio Barberini) who most probably provided financial assistance to the student.[34] Minerva, shown in the foreground, was usually represented in that period as the goddess who distributes the gifts of the mind. She is often represented as leading a prince or young man to the Arts. In Greuter's print she is seen presenting the young man, presumably the student, dressed probably as Aeneas. The student seems to recall the classical associations of the Barberini family with the travels of Aeneas recounted in the *Aeneid*. Aeneas landed in Lazio and pronounced it as his home, '*hic domus*', the Barberini family motto. Minerva is showing the student what is going to happen in the land he is about to colonise (the student's career), as if to say that after arduous study he is about to prosper under the protection and patronage of the Barberini family. The mountain in the centre perhaps refers to Mount Parnassus, following the Parnassus Plays (it is so identified in Hollstein's catalogue). The Parnassus Plays were a popular English satire of higher education published as a trilogy between 1598 and 1601 and became popular throughout Europe as a symbol of the frustrations of a student's four-year course. Here the mountain also clearly means Rome and the city of the Barberini popes, for the two old men in the left foreground certainly represent the Tiber. The kind of boat represented, commonly used to navigate up the Tiber, which is of the same style as that in Bernini's statue *La Barcaccia* in the Piazza di Spagna at the foot of the Spanish steps, confirms the interpretation.

The women at work represent happy *industria*. The bees (a pun on the arms of the Barberini) and honeycomb, are elements prescribed by Cesare Ripa in the *Iconologia*: women, bare feet, embroidery, and a swarm of bees.[35] Beneath the ground we see the Titans in the Underworld, from which the riches of industry are supposed to derive. Above ground, the women, who are being offered honeycombs, are embroidering maps: a nautical chart, a cosmographical world map representation on an oval projection, and a celestial globe, the three major realms of cartography of the sea, the earth, and the sky.

The maps are not only symbolic of knowledge and education, but also of industry, a meaning reinforced by the image of the map embroidering. The relationship between mapmaking and embroidery (weaving) as practical or productive arts was one that could be easily understood at the time. There is even a connection between embroidery (or weaving) and the use of a net of coordinates that we can clearly see in both the celestial globe and the world map, a theme echoed in several other contexts in the history of cartography.[36]

STUDY

Although the display value of prints and maps was becoming significant in the sixteenth century, it would be wrong to assume that this was their only function. Maps became an integral part of collections intended to provide a microcosm of universal knowledge, the *Wunderkammer* or *Kunstkammer*. Some amassed the goods of God's creation, the *naturalia* or natural history museum, such as the collection of Albrecht V and of Ferdinand II of Tirol at Schloss Ambras. Others dealt more with the *artificialia*, in what would be called today a museum of art or material culture. The Italian equivalents were the *studioli*, culminating in the celebrated one of Francesco I, second Grand Duke of Tuscany.[37]

In the late sixteenth century, it was usual for the great print collections of the nobility to contain topographical views and even maps among the albums devoted to the work of particular printmakers who had garnered a reputation for being collectible (such as Albrecht Dürer and Cornelis Cort). The collection of Cardinal Scipione Gonzaga, for example, contained a group of prints organised by subject, among which were probably views of ancient and modern Rome.[38] Similarly, the unusually well-documented *Kunstkammer* of the dukes of Bavaria contained a large number of prints, drawing, and paintings on parchment as well as maps (both mounted and unmounted) stored in drawers. And geography seems to have been a major interest of the Hapsburg Archduke Ernst, for his collection contains great quantities of topographical views, landscapes, maps, fortification prints, antiquarian views, and imaginary scenes, all apparently satisfying his need to understand a geographical context for contemporary and past events.[39]

Maps were not only the preserve of the ducal *Kunstkammer*, but also could be included in the more modest libraries and *studioli* of scholars and antiquarians. The printshops of Florence, Venice, and Rome became foci for discussion and intellectual stimulus. According to an eye-witness of the early 1550s, Tramezzino's and Salamanca's shops in Rome were a haunt of archaeologists and antiquarians interested in ancient Rome, and

Lafréry's shop in the via del Parione was a place to meet, discuss the latest ideas, and to add to one's collection.[40] Lafréry's pamphlet of his wares, published in 1573, extant in only one copy, consisted of almost 500 items.[41] Although maps and views were in the minority in Lafréry's business—they account for just over a fifth of the total—they were listed first in the catalogue. (Fig.33) Many of the maps can be positively identified, as with the first five maps in the catalogue: the world maps of Antonio Floriano, Forlani, Cimerlino, Tramezzino, and Forlani's engraved version of the chart of Europe and the Mediterranean by Diogo Homem.[42] Individual countries are then listed, starting in the north according to Ptolemy's order and ending in Asia. The New World forms a separate category, but is represented by only three maps: Cuba and Spagnola, Peru, and Nova Franza,[43] all maps that had previously been sold by Paolo Forlani in Venice, and for which the plates had very probably been brought by Claude Duchet when he left Venice in 1570.

There are scattered references to requests for Lafréry's wares. On 10 December 1563, Vincenzo Pinelli from Padua wrote to Paolo Manuzio in Rome to 'send him for inspection—with the price—a folio book of 150 pages of Roman antiquities, printed by Antonio Lafreri in Parione.'[44] Pinelli stressed that he wanted the 150-page, *not* the 300-page, version. This reference indicates a precursor of the *Speculum Romanae Magnificentiae* and is a clear indication that the market was beginning to mature in the 1560s.

Marc-Antoine Muret, who taught for 20 years in Rome and was a well-known antiquarian, wrote in a letter of 2 November 1572 to M. De Puy, lawyer at the Parliament, 'I talked with Lafreri these past few days. He sent me today everything he has done in a year and a half.'[45] Phillipus van Winghe wrote to Ortelius from Rome on 24 December 1589 reporting that he had bought the *Peregrinatio Pauli* in two sheets, after rummaging in a shop selling geographical maps. This reference is probably to Forlani's 1571 map of Europe showing the journeys of St. Paul, a reduction of Mercator's multi-sheet original of 1554.[46]

The functions of the *studiolo* could encompass both the function of private study and display. World maps and globes could

Indice
DELLE TAVOLE MODERNE
DI GEOGRAFIA DELLA
MAGGIOR PARTE DEL MONDO

di diuerſi auttor, iraccolte
& meſſe per ordine.

MAPPAMONDO in doi tondi per metter ſopra una palla.
Mappamondo ouero uniuerſale diſcrittione di tutta la terra.
Mappamōdo ouero Coſmografia uniuerſale in forma di cuore
Mappamondo di duoi pezzi in forma tonda.
Deſcrittione della nauigatione di tutta l' Europpa.

Frislandia
Heslandia } *Iſole ſettentrionali.*
Hibernia

Britania ouero Inghilterra
Nomi antichi & moderni in Anglia & Scotia
Vna parte d' Europpa
Spagna
Portogallo
Maiorica & Minorica Iſole
La Francia con li ſuoi confini
Vn altra deſcrittione della Francia
Ducato di Sauoia
Gallia Belgica
Deſcrittione della Fiandra
Barbantia
Geldria
Holandia
Friſia
Regno di Dennemarc con li ſuoi confini
Regno di Suenia, Nouergia & altre prouincie ſettentrionali
Germania

FIG. 33 Antonio Lafreri, [printed catalogue of his stock of maps and prints, Rome, 1573] fol. A2. Unique copy in Florence, Biblioteca Marucelliana, Misc.79.4.Cass.I.

also be used as scientific instruments as well as decorative objects of art hanging on the walls. The symbolic significance of the study as the head of household's private room was well established by the fifteenth century. The *studiolo* was a male preserve. In Alberti's treatise on family life, the narrator recalls: 'I kept my records at all times . . . locked up and arranged in order in my study, almost like sacred and religious objects. I never gave my wife permission to enter that place, with me or alone.'[47] Jacomo Contarini, the Venetian senator, mathematician, and instrument collector included a bequest 'for the perpetual conservation of his study and books' that began with the following statement: 'One of the most dearest things I have had, and have, is my studio, from which have come all personal honors and esteem.'[48]

A prime illustration of the valuable libraries amassed in Venice for which we have only tantalising references is that of Tommaso Rangone, already introduced as a patron of Paolo Forlani. About 1560, Tommaso opened in Venice 'a choice library for the use of scholars,' in which, in addition to printed books, there were many manuscripts in eastern languages, seals, cameos, medals, pictures of eminent men, mathematical instruments, astrolabes, spheres, celestial and terrestrial globes, 'geographical maps of learned authors,' and many other rarities.[49]

Sick with the plague in 1576, Tommaso bequeathed 500 ducats to enlarge his library for the public's use, but the books were given to the Capuccin of the Giudecca (Franciscan monks who officiated at the church of the Redentore). When the Franciscans' library was dispersed at the end of the 1700s, Tommaso's books and instruments disappeared with them. It is hard to imagine how rich this source would have been for our studies had it, or even a record of its contents, survived.

In Ulisse Aldrovandi's *studiolo* (called the room of mathematical instruments), there were not only models of bridges and astronomical instruments (quadrants, astrolabes, nocturnals, etc.) but also two globes (celestial and terrestrial), a small globe of the earth, a geographical map of the whole earth, and the *Italia nuova* of Giovanni Antonio Magini.[50] His visitors' book (1,579 names starting in 1566) included 1 pope, 10 cardinals, 21 bishops, 3 inquisitors, 907 scholars, 87 doctors, 1 cosmographer, and, perhaps surprisingly—considering earlier allusions to the sanctity of the study as a male preserve—1 woman. Among the visitors was Samuel Quiccheberg.

Quiccheberg held a post as physician at Albrecht V of Bavaria's court. After a period in Padua he seems to have taken up a position with Johann Jakob Fugger as a teacher and as the person responsible for the Fugger collection and library.[51] In 1565, he published a small handbook of 62 pages in four chapters on collecting, including what is arguably the earliest articulate plan for organising a map collection.[52] The title page reads: 'Rich theatre of objects of the whole universe, unique materials, and extraordinary representations.' In it, Quiccheberg explores the idea of an encyclopaedic text as encountered in Pliny's *Historia naturalis.* He makes the point that such a 'Wunderkammer on paper' would be within the reach not only of princes and scholars, but everyone. His concept of the 'universal theatre,' representing all fields of human endeavour, was perhaps the first treatise on museology. Quiccheberg goes on to identify five main categories (*classis*), each divided into ten or eleven subclasses (*inscriptiones*), and comments on the importance of a library within a collection, arranged according to a hierarchy of 'faculties' or subjects: theology, jurisprudence, mathematics, medicine, and 'museum studies.' Adamant that the collections of images should be separate and important in their own right, he advises that loose sheets should be kept between covers and clearly titled. The treatise also advocates binding groups of engravings that appear complete and recommends handling them as books and demands the systematic cataloguing of collections.

Maps had a prominent place in these collections of engravings.[53] They are included in two sections: the fourth to the tenth *inscriptiones* of the first *classis*, where they appear as a topographical context for the collector's family and local history; and in the fifth *classis*, where they seem to be part of a collection of works of art, again glorifying the collector's position in society, but prized for their artistic technique rather than their topographical content.

It is difficult to assess the impact of Quiccheberg's volume because it is not acknowledged in later manuals of museum collecting.[54] But his scheme was obviously influential in forming the Albrecht collection and other ducal collections in the late sixteenth century in the movement towards fully-fledged museums.

TOPICAL INFORMATION

The dissemination of news to the public relied in good part on the services of the *venditori ambulanti* in the sixteenth century. In *La Cortigiana*, Pietro Aretino paints a vivid picture of Furfante, the Roman street peddler:

> Pretty stories, stories, stories, the war of the Turks in Hungary, the sermons of Father Martino, the Council, stories, stories, the facts of England, the festivities of the Pope and the Emperor, the circumcision of the Voivoda, the Sack of Rome, the Siege of Florence, the skirmish at Marseilles and its conclusion, stories, stories . . .'[55]

Maps and prints were sold along with such news stories, and the information overload we normally associate with the late twentieth century was clearly not unknown in the sixteenth century. In the print illustrated in Fig.34, the seller, replete with a portrait of a Turk, a view of Buda, a map of central Europe (Allemania), portraits, and broadsides, some of which are pinned to his clothes, is shouting: 'let him who wants buy—news of war, maps of war, cheap at two bolognini each!!' His hearers are not so sanguine. The one on the right is saying: 'we're full of news, go away, go away, we're fine as we are.' The one on the left is saying: 'I don't want to hear any more news, no, no, no!'[56]

It is important for us to understand how people five hundred years ago could have been informed about crucial events taking place hundreds of miles away and separated by sea. For example, the Siege of Malta generated an extraordinary press in 1565, documenting step by step how a few hundred Knights of Malta successfully withstood a siege by the Ottoman army for almost four months until relieved by long awaited reinforcements. To the

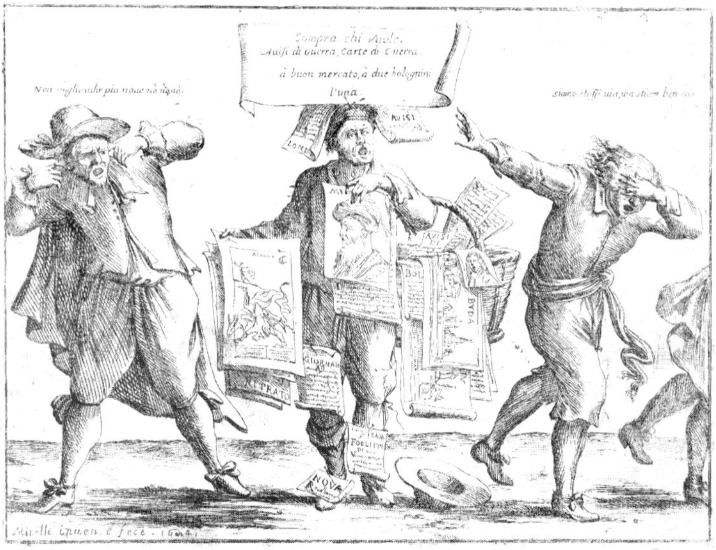

FIG.34 G. M. Mitelli, [itinerant printseller of war maps and prints], 1688. BM. P&D 18 52-10-9-1041.

ordinary sixteenth-century citizen, the fate of Europe at the hands of the Ottomans was perceived to hinge on its outcome. Even quite out-of-date atlases were annotated with the momentous event. For example, a copy of the 1511 Venetian edition of Ptolemy's *Geography* has a manuscript annotation to the effect that 'on 18 May 1565 the Turkish armada arrived with 130 galleys and 12 ships but after having left over 20,000 men dead, left ignominiously.' The owner of the Ptolemy had a more recent map of Malta bound in the back of the atlas.[57]

The first graphic news of the siege, in the form of Niccolò Nelli's map, appeared only a month after the first events of the siege, engraving and printing time included. News of the final Turkish retreat took only six days to reach Rome, and only eleven more to reach Brussels. (Fig.35) The base for Nelli's maps can be traced to a drawing by Bartolomeo Genga, an Italian mil-

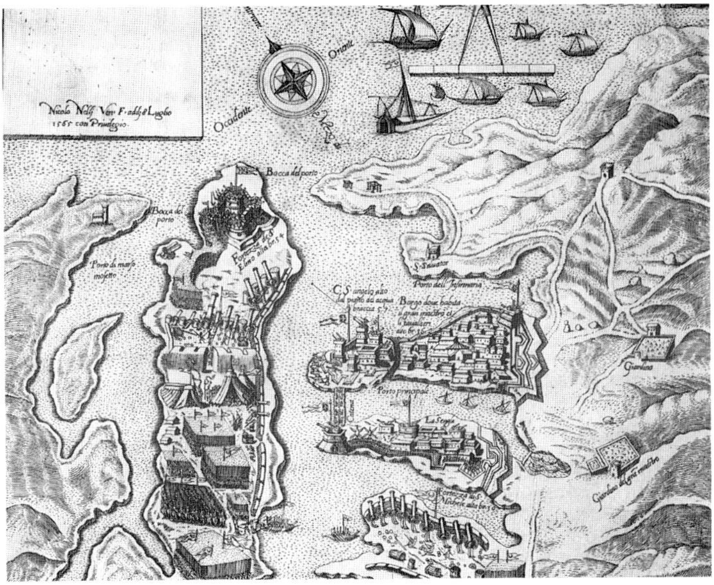

FIG.35 Nicolò Nelli, *Il porto dell'Isola di Malta*, 8 July 1565 state 1. Ganado 2. Milan Civica Raccolta di Stampe Bertarelli EE.46.216.

itary engineer who, shortly before his death in 1558, submitted to the Grand Master Jean de la Valette a design for the construction of a fortified city between the two harbours.[58]

The map shows the first stage of the siege nearing its end. One main line of guns is trained on Fort St. Elmo. At the height of the siege of St. Elmo on 19 June, there had been 34 pieces bombarding the fort. Although St. Elmo is still shown to be resisting (the flag of the Knights of Malta is still flying), there are features shown that relate to the second stage of the siege after the capture of the fort on 23 June. These include the start of operations against Senglea and Fort St. Michael, represented by nine guns. Although the map was probably not on the market by the date shown (8 July), it was certainly available later in the same month.

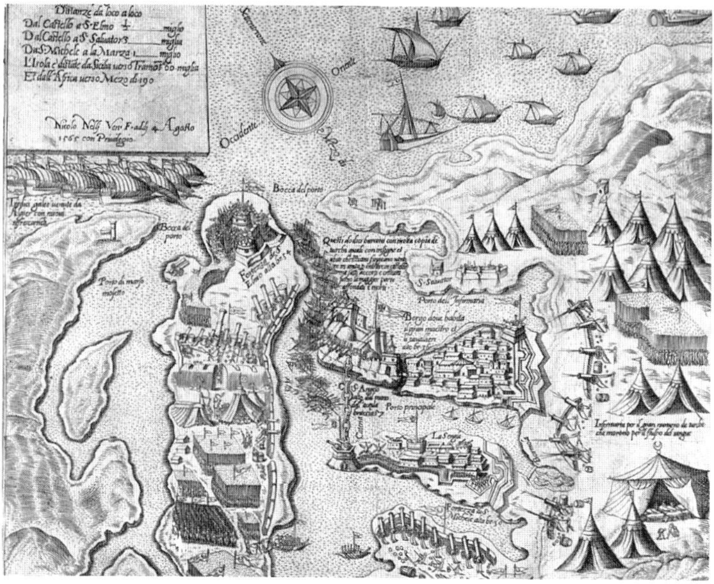

FIG.36 Nicolò Nelli, *Il porto dell'Isola di Malta*, 4 August 1565. state 2. Ganado 3. Director of Museums, Malta.

The next state of the Nelli plate is dated 4 August 1565. (Fig.36) The publisher brought out three main events: the arrival off Sliema Point of 2500 Ottoman reinforcements from Algiers led by Hassan Agà or Bey late on Sunday 8 July; a feint sea assault on Fort St. Angelo (an event not noted in any of the usual textual sources); and new batteries installed against Borgo and St. Michael. News of the sea assault on Senglea (mid-July) had not reached Nelli, and St. Elmo is still shown flying the Knights' flag. However, the words 'presa del forte' (capture of the fort) are shown.

On the next version, there were no substantive changes, but spaces were left blank for the day to be written in manuscript. Nelli had now become aware that changes would be needed every few days, and it was going to be more efficient to fill them in by hand. (Two versions, the 5 and 12 August, are known).

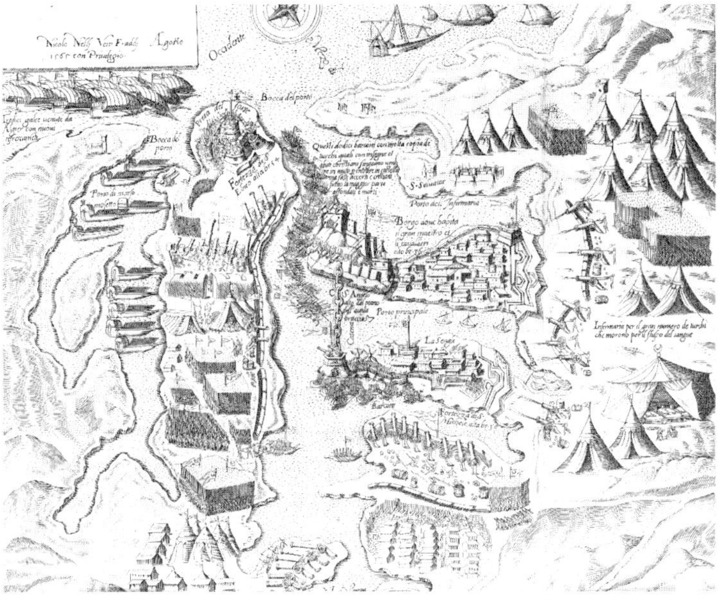

FIG.37 Nicolò Nelli], *Il porto dell'Isola di Malta*, August 1565. state 5 Ganado 6. BL 9150.b.12.

While not a 'state' of the copperplate in the normal cartobibliographic sense, the systematic issue of impressions annotated with the same manuscript date surely constitutes a bibliographic practice for which a term is needed. On the next state, probably issued in late August (with the blank for the day still showing), ten Turkish galleys are shown safely at anchor inside Marsamuscetto, unthreatened by the conquered Fort St. Elmo. (Fig.37) The prime motive for taking the fort was to assure this anchorage. The abortive assault on Senglea is shown. Under 'Agosto' the erasure of the date 8 Luglio from the previous state is discernible as a ghost print resulting from the wearing down of the plate.

When issued undated by Claude Duchet, probably in the late 1570s, the plate underwent little substantive change, but had

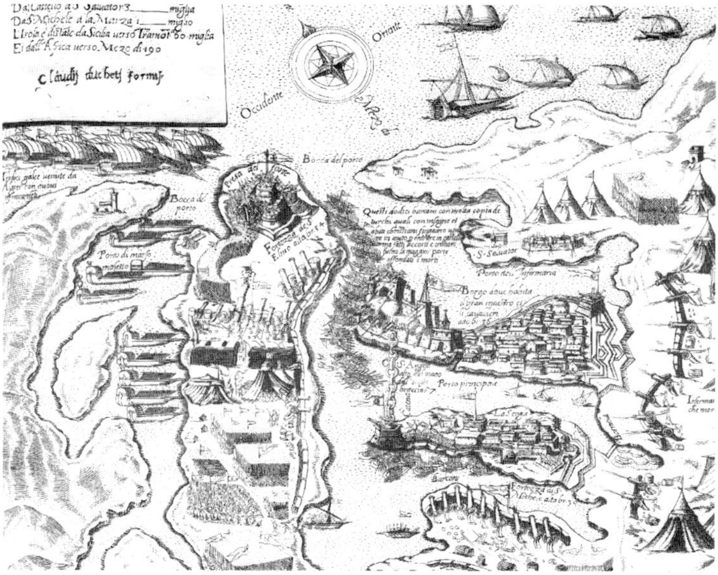

FIG.38 [Nicolò Nelli, *Il porto dell'Isola di Malta* [Rome], Claudio Duchetti, [1565-1585] state 6. Ganado 7. San Marino, California, Huntington Library (LAF 100).

undergone considerable wear. (Fig.38) The basilisk on Mount Sceberras has opened fire and heavy hatching has been added to the gun platforms.

On a different plate, produced by Nelli in October 1565 from an original by Giovanni Battista Pittoni, the Turks are seen rapidly retreating into their galleys and sailing away. (Fig.39) The final battle had taken place on 11 September. Four days before, the long awaited relief force under Don Garcia had arrived in northwest Malta. When word reached the Turkish commander, he ordered an immediate embarkation, but returned to St. Paul's Bay just to the west and marched to the capital Mdina. There on 11 September, they were routed by mostly Spanish troops under Chiappino Vitelli and were chased into the sea at St. Paul's Bay

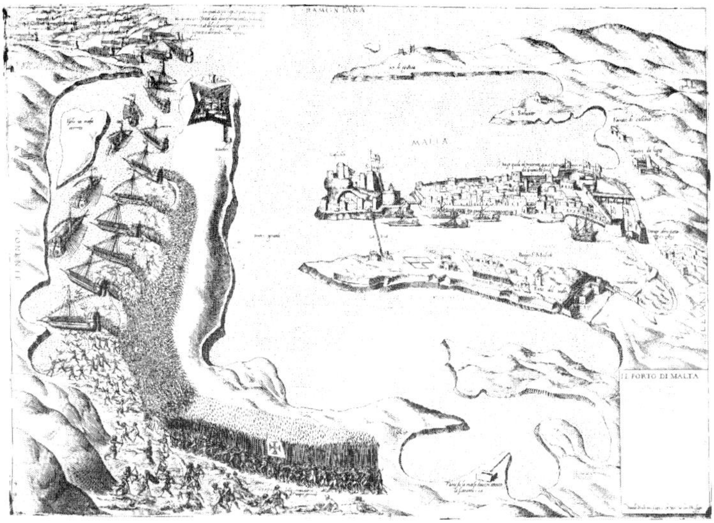

FIG.39 [Giovanni Battista Pittoni] Malta, *Il porto di Malta* [October] 1565 [Venice], Nicolò Nelli, 1565. State 1. Ganado 8. BL Map Library 24340(2).

where the galleys were waiting. The fleet lay offshore and four hours before sunrise on 13 September started on the long trip back home. In the second state was added a portrait of Jean de la Valette, the Grand Master who against all odds had held out the siege in Borgo for two months from the date reinforcements had been requested on 8 July.

The speed of circulation and impact of these news documents in sixteenth-century Europe should not be underestimated. Neither should the longevity and durability of these broadsheet images: there were over 140 separate representations of the encounter over the next eighty years.[59]

CONCLUSIONS

Printed maps were regarded as geographical prints in the Italian Renaissance. They were engraved by the same engravers, made in the same printshops, sold by the same street sellers, appeared on both sides of the same sheet of paper, and were gathered within the same bound volumes with other prints. Maps were subject to the same technical and marketing constraints as other figurative prints and were designed to be sold to an expanding market of people eager for news or wanting to collect and display maps to enhance their reputation as cosmopolitan citizens.

The role of maps as decorative items in forming the taste for a new consumerism should not be denigrated. The consumer's viewpoint offers the historian insights into the social conditions in which maps and prints acquired meaning and value, whether in an aesthetic or intellectual sense.

It was the ability of the figurative print to disseminate and maintain a standard syntax of representation that partly accounted for their efficiency in conveying information. Such standardisation had been achieved through the development of systematic hatchings to show the third dimension, a trait important for both maps and figurative prints. The range of graphic printed matter was staggering: engineering drawings or pattern books, herbal or anatomical manuals, emblem books, fortification and hydrological drawings, and maps and plans of all kinds infused the culture with useful information. These new kinds of figurative prints were not as numerous as traditional religious pictures in the sixteenth century, but they provided the natural sciences with the ability to compare and classify in 'paper universal theatres,' replacing memory in a permanent form and transmitting information from one individual to the next without much loss of accuracy. The figurative print thus laid the groundwork for the taxonomic breakthroughs of the following century.

The wares of the printshops of Venice and Rome might not have been avant-garde in disseminating new information about geographical discoveries, since most of this was in the form of closely controlled official manuscripts. However we should not conclude from this that these largely derivative maps had no useful effect on the spread of geographical information or forming world views. As a source of topical and informal information, the geographical print supplied the general public with information about current events such as the Siege of Malta with staggering immediacy. The details might not always be quite correct, but in some cases the maps provide us with information not available from other sources.

That there was a radical change in the patterns of ownership of maps and works of art between the fifteenth and sixteenth centuries is not in doubt. Clearly, by the mid-1560s, the printed map trade had matured and was driven by a market substantial enough to demand bound sets of maps and prints. The wider dissemination of printed images had probably affected the use of maps for practical administration much earlier, particularly in Florence, where the short-lived trade had already been established at the turn of the sixteenth century. It is difficult, considering the evidence for map ownership in the early sixteenth century, to argue for the view that 'Mapping as a normal administrative way of looking at the world dates only from the third quarter of the sixteenth century'. It seems clear that mapping was not an *abnormal* administrative way of looking at the world at the end of the *fifteenth* century, and that evidence for considerable social uses of maps—the incipient map trade—is present for the first half of the sixteenth.

Recent approaches to map history have proposed an alternative epistemology, based not on scientific positivism but on social theory, 'to search for the social forces that have structured cartography and to locate the presence of power–and its effects–in all map knowledge.' We can agree with the value of reading 'between the lines of the map . . . to discover the silences and contradictions that challenge the apparent honesty of the image.'[60] Indeed, the subject of cartographic ethics has recently seen a revival of interest as it becomes clear how messages can be blatantly manipulated to serve the ends of the compiler.

We can now extend the discussion of the presence of power in maps to the symbolic roles of maps in shaping social tastes and world views. Here the ideology of individual makers and agencies in moulding the information content of maps is less important than the cultural practices that constitute the production, distribution, and consumption of symbolic goods. In other words, maps not only were viewed as sources of geographical information but also as displayed symbols of social status. The print workshops of Renaissance Florence, Venice, and Rome were in the vanguard in producing an impressive range of geographical prints that acquired meaning and value in the everyday life of Early Modern European society. It is by looking at maps in the context of the book trade and print trade that the importance of the consumer emerges.

1 K. A. Manley, 'The Book Wolf Bites a Bohn: Panizzi, Henry Bohn and Legal Deposit, 1850–53', In *The Library of the British Museum: Retrospective Essays on the Department of Printed Books*, ed. P. R. Harris. 145–64. (London: The British Library, 1991), 147.
2 Manley, 163.
3 A notable exception is Elizabeth L. Eisenstein, *The Printing Press as an Agent of Change* (Cambridge: Cambridge University Press, 1979), which fully treats the impact of graphic printing as well as typographic printing.
4 Recent issues of *Imago Mundi: The International Journal for the History of Cartography*, vols. 46–48, have been especially rich in these studies.
5 Antonio de Beatis, 'The Travel Journal of Antonio de Beatis: Germany, Switzerland, The Low Countries, France and Italy, 1517–1518', *The Hakluyt Society Second Series No. 150*, (London: Hakluyt Society, 1979).
6 Allan W. Atlas, 'Aggio visto lo mappamondo: A New Reconstruction', In *Studies in Musical Sources and Style: Essays in Honor of Jan LaRue*, ed. Eugene K. Wolf and Edward H. Roesner, (Madison, Wisconsin: A-R Editions, 1990), 109–17.
7 Marino Sanuto, *I diarii*, ed. Rinaldo Fulin, (Bologna: Forni, 1969), III, viii, col. 247.
8 David Chambers, Brian Pullan, and Jennifer Fletcher, eds., *Venice: A Documentary History, 1450–1630*, (Oxford: Blackwell, 1992), 405, citing a letter of Isabella d'Este to her husband Francesco Gonzaga, 10 November 1506.
9 David Landau and Peter Parshall, *The Renaissance Print: 1470–1550* (New Haven and London: Yale University Press, 1994), 13. The Rosselli inventory was first published by Jodoco Del Badia, 'La bottega di Alessandro di Francesco Rosselli merciaio e stampatore', *Miscellanea fiorentina di erudizione e di storia* 2 (1894), 24–30; See also Roberto Almagià, 'On the Cartographic Work of Francesco Rosselli', *Imago Mundi* 8 (1951), 26–34; Jay A. Levenson, Konrad Oberhuber, and Jacquelyn L. Sheehan, *Early Italian Engravings from The National Gallery of Art*, (Washington, D. C.: National Gallery of Art, 1973); and John Goldsmith Phillips, *Early Florentine Designers and Engravers* (Cambridge: Harvard University Press, 1955).
10 Roberto Almagià, 'Il primato di Firenze negli studi geografici durante i secoli XV e XVI', In *Atti della Società Italiana per il Progresso delle Scienze* (Florence, 1929); Sebastiano Gentile, *Firenze e la scoperta dell'America: Umanesimo e geografia nel '400 Fiorentino* (Florence: Leo S. Olschki, 1992).
11 Leonardus Vytautas Gerulaitis, *Printing and Publishing in Fifteenth-Century Venice* (Chicago: American Library Association, 1976).
12 Paolo Veneziani, 'Vicende tipografiche della Geografia di Francesco Berlinghieri', *La Bibliofilia* 84 (1982), 195–208.

13 Claudius Ptolemy, *The Geography*, translated and edited by Edward Luther Stevenson, with an introduction by Prof. Joseph Fischer, S.J. (New York: New York Public Library, 1932; reprinted Dover Publications, 1991). On these distinctions, see Denis Cosgrove, 'Mapping New Worlds: Culture and Cartography in Sixteenth-Century Venice', *Imago Mundi* 44 (1992), 65-89; John F. Moffitt, Medieval *Mappaemundi* and Ptolemy's *Chorographia*', *Gesta* 32 (1993), 59-68; Thomas Frangenburg, 'Chorographies of Florence: The Use of City Views and City Plans in the Sixteenth Century', *Imago Mundi* 46 (1994), 54 and 56 and fig.6; Lucia Nuti, 'The Perspective Plan in the Sixteenth Century: The Invention of a Representational Language', *Art Bulletin* 76 (1994), 105-28.

14 Early editions of Ptolemy's *Geography* contain the word *cosmography* in their incipits or colophons following Jacopus Angelus's original translation from the Greek in 1409. In the 1478 Rome edition, Ptolemy is referred to as 'philosophi geographiam' in the colophon, and the Berlinghieri edition of 1482 contains a 'title' that includes 'Geographia.' This trend of calling the book the Geography is continued in the 1490 Rome edition and *all* subsequent editions of Ptolemy, while the sixteenth-century authors of geographical works tended to entitle their works 'cosmographies'.

15 Giacomo Gastaldi, *La universale descrittione del mondo, descritta da Giacomo de' Castaldi Piamontese* (Venice: Matteo Pagano, 1561), 3-4.

16 Francesco Ferretti, *Diporti Notturni. Dialloghi familliari del Cap*^{o}. *Franc*^{o}. *Ferretti cav*^{ro}. *dell'ordine di Sa*^{to} *Stefano con la dimostratione figurale intagliata da Michelangelo Marelli Anconitano* (Ancona: Francesco Salvioni, 1580).

17 Accademia della Crusca, *Vocabolario degli accademici della Crusca* (Florence: Tip. Galileiana di M. Cellini, 1863-1923).

18 Frangenburg, 52-55.

19 *Italia Nuova*, 1558, Tooley 327. Tooley numbers refer to the catalogue in Ronald Vere Tooley, Maps in Italian Atlases of the Sixteenth Century, Being a Comparative List of the Italian Maps issued by Lafreri, Forlani, Duchetti, Bertelli and Others, Found in Atlases', *Imago Mundi* 3 (1939), 12-47.

20 Robert W. Karrow, Jr., *Mapmakers of the Sixteenth Century and Their Maps: Bio-Bibliographies of the Cartographers of Abraham Ortelius, 1570* (Chicago: Speculum Orbis Press for The Newberry Library, 1993).

21 Leonardo Rombai, 'La formazione del cartografo nella Toscana moderna e i linguaggi della carta', In *Imago et descriptio Tusciae: La Toscana nella cartografia dal XV al XIX secolo*, ed. Leonardo Rombai, (Florence: Giunta regionale toscana, 1993), 37-81.

22 Germaine Aujac, 'Le peintre florentin Piero del Massaio et la *Cosmographia de Ptolémée*', *Geographia Antiqua* 3-4 (1994-95), 187-210. The manuscripts are Rome, Vatican Library, Vat. Lat. 5699 [1469], Vat. Urb. 277 [1472-73], and Paris, Bibliothèque Nationale, Par. Lat. 4802 [ca. 1474-75].

23 Margherita Azzari, 'Vedutismo pittorico e cartografia locale nella Toscana del Quattrocento', In *Il mondo di Vespucci e Verrazzano: geografia e viaggi*, ed. Leonardo Rombai. (Florence: Leo S. Olschki, 1993).

24 Leonardo Rombai, 'La nascita e lo sviluppo della cartografia a Firenze e nella Toscana granducale', In *Imago et descriptio Tusciae: La Toscana nella cartografia dal XV al XIX secolo*, ed. Leonardo Rombai, (Florence: Giunta regionale toscana, 1993), 91.

25 Mario Signori, 'Ingegneri e idrografia: le carte e le acque', In *L'Europa delle carte: Dal XV al XIX secolo, autoritratti di un Continente*, ed. Marica Milanesi, (Milan: Gabriele Mazzotta, 1990), 49–52.

26 Azzari, 97.

27 Signori, 50.

28 Thomas Ashby, *La Campagna Romana al tempo di Paolo III. Mappa della Campagna romana del 1547 di Eufrosino della Volpaia riprodotta dall'unico esemplare esistente nella Biblioteca Vaticana a cura della Biblioteca medesima* (Rome: 1914).

29 Rombai, 'Formazione del cartografo', 43.

30 Florence, Archivio di Stato di Firenze, Carte nautiche, n.13.

31 Geography. *Modern maps of most of the world by various authors collected and put in Ptolemy's order with views of many cities and forts from various regions. Published from copper plates with care and diligence in Rome.*

32 Respectively Tooley 28, 64, 71 and 75 (Fig.5), 179, 338, 287 and 453.

33 Cosgrove, 75.

34 Michel Foucault, 'What Is an Author?', In *Textual strategies : perspectives in post-structuralist criticism*, ed. Josue V. Harari, (Ithaca, N.Y.: Cornell University Press, 1979), 141–60.

35 Tony Campbell, *The Earliest Printed Maps* (London: British Library Publications, 1988).

36 David Woodward, 'Medieval Mappaemundi', In *History of Cartography: Cartography in Prehistoric, Ancient, and Medieval Europe and the Mediterranean*, ed. J. B. Harley and David Woodward, (Chicago: University of Chicago Press, 1987), I, 286–370.

37 M. Maylender, *Storia delle Accademie d'Italia*, 5 vols., (Bologna: Licinio Cappelli, 1926–30, repr. Bologna, Forni, n.d.). For more recent discussions, see Martin Lowry, 'The Proving Ground: Venetian Academies of the Fifteenth and Sixteenth Centuries', In *The Fairest Flower: The Emergence of Linguistic National Consciousness in Renaissance Europe*, (Florence: 1985), 41–51 and Gino Benzoni, Per un profilo dell'Italia accademica', *Atti dell'Istituto Veneto di scienze, lettere ed arte: Classe di scienze morali, lettere ed arti* 152 (1994), 1–44.

38 Paul Lawrence Rose, 'The Accademia Venetiana: Science and Culture in Renaissance Venice', *Studi veneziani* 11 (1969), 191–242, esp. 194.

39 Rose, 199.

40 Rose, 193.
41 Rose, 193.
42 Karrow, 216–49.
43 David Woodward, 'La Geografia Moderna: The Work of the Sixteenth-Century Italian Cartographers', *Mapline* Special number 4 (1979).
44 Any undiscovered islands were of particular interest for the spreading of the Catholic faith and in dividing the world between Spain and Portugal following Columbus's first voyage. See the transcription of Alexander VI's papal bull 'Inter Caetera' of 4 May 1493 in Agustìn Remesal, *1494: La Raya de Tordesillas* (Junta de Castilla y Leon, 1994), 133–37.
45 Baldasar Castiglione, *The Book of the Courtier*, ed. George Bull, (London: Penguin Books, 1967), 96–97.
46 Quoted in Eva Mabel Tenison, *Noble arts, especially maps: Notes on hitherto unknown examples of sixteenth century cartography. With annotated lists of maps, charts, documents, and pictures, published in portfolio accompanying the eight quarto volumes of Elizabethan England* (Glasgow: Glasgow University Press, for the Author, 1932).
47 G. Fanelli, *Firenze* (Bari: 1980), 77.
48 Peter W. Parshall, 'Imago Contrafacta: Images and Facts in the Northern Renaissance', *Art History* 16 (1992), 554–79.
49 Lucia Nuti, 'The Perspective Plan in the Sixteenth Century: The Invention of a Representational Language', *Art Bulletin* 76 (1994), 115.
50 Francesco Ehrle, *Roma prima di Sisto V: La pianta di Roma Du Pérac-Lafréry del 1577 riprodotta dall'esemplare esistente nel Museo Britannico. Contributo alla storia del commercio delle stampe a Roma nel secolo 16 e 17* (Rome: Danesi, 1908). See also Paolo Bellini, 'Stampatori e mercanti di stampe in Italia nei secoli XVI e XVII', *I quaderni del conoscitore di stampe 26* (1975), 19–22, translated as 'Printmakers and Dealers in Italy during the 16th and 17th Centuries', *Print Collector* 13 (1975), 17–45; Laura Biancini and Ludovica Mazzola, ed., *Guida alla catalogazione per autori delle stampe*, (Rome: Istituto Centrale per il Catalogo Unico, 1986); and David Woodward, 'Paolo Forlani: Compiler, Engraver, Printer, or Publisher?', *Imago Mundi* 44 (1992), 45–64. For an excellent source of information and literture on the seventeenth-century Parisian print trade, see Marianne Grivel, *Le commerce de l'estampe a Paris au XVIIe siècle* (Geneva: Librairie Droz, 1986).
51 David Woodward, 'The Forlani Map of North America', *Imago Mundi* 46 (1994), 29–40.
52 Del Badia, 1894 and the reprinting with notes in Arthur Mayger Hind, *Early Italian Engraving. A Critical Catalogue with Complete Reproduction of all the Prints Described* (London: 1938–48), I, 304–09.
53 See particularly the transcription of the account book by Francesco Calzolari, Jr: Conor Fahy, *Printing a Book at Verona in 1622* (Paris: Fondation Custodia, 1993), and Francesca Consagra, 'The De Rossi Family Print

Publishing Shop: A Study in the History of the Print Industry in Seventeenth-Century Rome', Ph.D. Dissertation, The Johns Hopkins University, 1993 and her article 'De Rossi and Falda: A Successful Collaboration in the Print Industry of Seventeenth-Century Rome', In *The Craft of Art: Originality and Industry in the Italian Renaissance and Baroque Workshop*, ed. Andrew Ladis and Carolyn Wood, (Athens: University of Georgia Press, 1995), 187–203. The Rosselli inventory is well described in Landau and Parshall, *Renaissance Print*, 12–13. A new, apparently fuller, manuscript of the Rosselli inventory was recently found, but a close comparison has yet to my knowledge to appear in print. See Gentile, *Firenze*, 247–50. For comments on the scarcity and obscurity of archival sources in the book trade of this period, see Fahy, *Verona*, 32–33.

54 Richard A. Goldthwaite, *The Building of Renaissance Florence: An Economic and Social History* (Baltimore and London: 1980), Appendix 3. An explanation of approximate currency equivalents is offered in the present volume as Appendix 1.

55 Goldthwaite, 347.

56 Jan de Vries, 'Between Purchasing Power and the World of Goods: Understanding the Household Economy in Early Modern Europe', In *Consumption and the World of Goods*, ed. John Brewer and Roy Porter, (London: Routledge, 1993), 89.

57 Richard MacKenney, *Tradesmen and Traders: The World of the Guilds in Venice and Europe, c.1250-c.1650* (London: Croom Helm, 1987), 98.

58 Consagra, 402–03.

59 Fahy, *Verona in 1622*, 35–37.

60 Gian Ludovico Masetti Zannini, *Stampatori e librai a Roma nella seconda metà del Cinquecento: Documenti inediti* (Rome: Fratelli Palombi Editori, 1980), 212.

61 MacKenney, [18].

62 David Woodward, *The Maps and Prints of Paolo Forlani: A Descriptive Bibliography* (Chicago: The Newberry Library, 1990).

63 Abraham Bosse, *Traicté des manières de graver en taille douce sur l'airain, par le moyen des eaues fortes et des vernix durs et mols* (Paris: 1645); Furio de Denaro, ed., *Domenico Tempesti e I discorsi sopra l'intaglio ed ogni sorte d'intagliare in rame da lui provate e osservate dai più grand'huomini di tale professione* (Florence: Studio Per Edizioni Scelte, 1994), 147.

64 Woodward, Forlani: Compiler, Engraver, Printer, or Publisher', 51.

65 Jan de Vries estimates the work year in the fifteenth century in the Netherlands as 250 days, comparable to Goldthwaite's estimate of 270 days for Italy. In Italy, the number of feast days is unlikely to have decreased in the sixteenth century as much as the Netherlands.

66 Consagra, 159. 1 libbra = 327.45 grams or about the equivalent of .723 modern lb.

67 Figures for the weight of copper plates need to be systematically compiled, but one would not necessarily expect a straight line relationship between surface area and weight. Seven copper plates ranging in area from 4231 cm^2 to 5374 cm^2 for the *Atlantic Neptune* were weighed and compared by Ed Dahl, National Archives of Canada, producing a range of 589 cm^2/kg to 773 cm^2/kg, suggesting that at least for similar sized plates, there is no correlation for size and weight, suggesting that thickness was irregular. For further background on the *Atlantic Neptune* plates, see Christopher Terrell, A Sequel to the *Atlantic Neptune* of J. F. W. DesBarres: The Story of the Copperplates', *Map Collector* 72 (1995), 2–8.

68 At the end of the fifteenth century there was a growing demand for copper by the shipbuilding industry that intensified in the early sixteenth century. The first wave in high copper prices came between 1527 and 1535 and the price continued to rise throughout the sixteenth century, doubling between 1530 and 1560 and doubling again between 1560 and 1620. The Augsburg merchants controlled 60–65% of the market and the Nuremberg companies held 27–33%. See Ekkehard Westermann, Tendencies in the European Copper Market in the 15th and 16th Centuries', in *Precious Metals in the Age of Expansion*, ed. Hermann Kellenbenz, (Stuttgart: Klett-Cotta, 1981), 71–77.

69 Vannoccio Biringucci, *De la pirotechnia: libri x* (Venice: V. Rossinello, 1540).

70 Eva Hanebutt-Benz, 'Copperplate Engraving and Printing', In *Kartographie und Druckkunst um 1500* (Mainz: International Map Collectors Society Meeting at the Gutenberg-Museum, September 1993), unpaginated typescript. The source of this information, which has also been cited elsewhere, is unknown. Extant copper plates have been an underutilised source of technical information, despite their wide availability, for example, in the State Archives of Turin and the Calcografia Nazionale, Rome. See Barbara Bertini Casadio and Isabella Ricci Massabò, *I rami incisi dell'Archivio di Corte: Sovrani, battaglie, architetture, topografia a cura dell'Archivio di Stato di Torino* (Turin: Archivio di Stato di Torino, 1982) and Carlo Alberto Petrucci, *Catalogo Generale delle stampe tratte dai rami incisi posseduti dalla Calcografia Nazionale* (Rome: Libreria dello Stato, 1953).

71 The problems of establishing sizes of paper and copper plates in the Rosselli inventory are summarised by Landau and Parshall, 16, in which it is shown that often quoted 1398 Bolognese dimensions of *reale* (44.5 × 61.5 cm) are probably twice the area of those used in the sixteenth century (29.5 × 43.4 cm). This is confirmed by the entry in Rosselli's inventory for the copper plates for a 'appamondo a mantelino, 2 fogli reali in tutto' which no doubt refers to the world map on a conic projection by Giovanni Matteo Contarini, engraved by Francesco Rosselli in 1506 (BL Maps C.2.cc.4), whose two sheets each measure about 31 × 42 cm.

72 De Denaro, 144.
73 Woodward, Forlani Map of North America', 00.
74 De Denaro, 148.
75 De Denaro, 148: 'There are other [stones] which come from Liège which are naturally soft and white, but anyone who has a good stone from the Levant doesn't need any other'.
76 Consagra, 302.
77 Consagra, 158.
78 Fahy, *Verona*, calculated from the transcription of the account book, 91-124.
79 Fahy, *Verona*, 57 and 59. A similar amount seems to have been owed by Ulisse Aldrovandi to the heirs of Francesco de Franceschi Senese in 1599 for engraving the frontispiece of his *Ornithologia* (£52 Bolognese for half). See *Intorno alla vita e alle opere di Ulisse Aldrovandi: studi* (Bologna, 1907), 82.
80 Consagra, 399.
81 Consagra, 398.
82 Antonino Bertolotti, *Artisti veneti in Roma nei secoli XV, XVI e XVII: Studi e ricerche negli archivi romani* (Bologna: Arnaldo Forni, 1884), 35. Bertolotti published a series of books between 1877 and 1889 listing archival references to artists of the following origins working in Rome: Sicily; Belgium and Holland (with additions), Urbino; Lombardy (with additions); Modena, Parma, and Lunigiana; Venice; Subalpine; Switzerland; Bologna, Ferrara and other regions in the Papal States; France; Germany; Sardinia, Sicily, Corsica, and Malta.
83 Woodward, *Forlani Bibliography*, iii.
84 *Fiorenza*, Claude Duchet, n.d., Tooley 206.
85 Marcel Destombes, 'Les cartes de Lafreri et assimilées (1532-1586) du Département des Estampes de la Bibliothèque Nationale', *Nouvelles de l'Estampe* 5 (1970), 234-74.
86 Bosse, [00.]
87 See Hind, *Early Italian Engraving*, nos. 19-22, 170-71 and plates 734-35; Pietro Manzi, 'Alcuni documenti di cartografia nolana ovvero: Ambrogio Leone e Gerolamo Moceto', *L'Universo* 53 (1973), 811-18.
88 David Woodward, 'The Manuscript, Engraved, and Typographic Traditions of Map Lettering', In *Art and Cartography: Six Historical Essays*, ed. David Woodward (Chicago: University of Chicago Press, 1987), 174-212.
89 Michael Bury, *Giulio Sanuto: A Venetian Engraver of the Sixteenth Century* (Edinburgh: National Gallery of Scotland, 1990).
90 Tony Campbell, 'Letter Punches: A Little Known Feature of Early Printed Maps', *Print Quarterly* 4 (1987), 151-54 and Campbell, *Earliest Printed Maps*, Appendix 2, 223-25.

91 Landau and Parshall, 27-28.

92 Sue Welsh Reed and Richard Wallace, eds., *Italian Etchers of the Renaissance & Baroque* (Boston: Museum of Fine Arts, 1989), xix, quoting Benvenuto Cellini and Giorgio Vasari on the subject in 1568.

93 Vittorio Zonca, *Novo teatro di machine et edificii per varie et sicure operationi. Co[n] le loro figure tagliate in rame e la dichiaratione, e dimostratione di ciascuna. Opera necesaria ad architetti, e a quelli, chi di tale studio si dilettano* (Padua: P. Bertelli, 1607). Reprint edited by Carlo Ponti, (Milan: Edizioni il Polifilo, 1985).

94 See Landau and Parshall, 324-32, for an excellent overview of the early history of etching.

95 Woodward, Forlani, Compiler, Engraver, Printer, or Publisher?' 48.

96 David Woodward, 'The Woodcut Technique', In *Five Centuries of Map Printing*, ed. David Woodward. (Chicago: University of Chicago Press, 1975), 25-50.

97 'Blockcutter' seems to be the closest English equivalent of the German *Formschneider* since *xylographer*, the English equivalent of the Italian *silografo*, is not commonly used.

98 Jacques Kuhnmünch, 'Le commerce de la gravure à Paris et à Rome au XVIIe siècle', *Nouvelles de l'estampe* 55 (1981), 6-17, 6.

99 Leo Bagrow, *Giovanni Andreas di Vavassore: A Venetian Cartographer of the 16th Century. A Descriptive List of his Maps* (Jenkintown, PA: George H. Beans Library, 1939); George H. Beans, 'Some Notes from the Tall Tree Library: Vavassore and Pagano', *Imago Mundi* 5 (1948), 73; George H. Beans, 'A Note from the Tall Tree Library (Vavassore maps)', *Imago Mundi* 10 (1953), 14.

100 Charles Ephroussi, 'Zoan Andrea et ses homonymes', *Gazette des Beaux-Arts* 3d ser., 5 (1890), 401-15, and 6 (1891), 225-44.

101 Enid T. Falaschi, 'Valvassori's 1553 Illustrations of *Orlando furioso*: The Development of Multi-Narrative Technique in Venice and Its Links with Cartography', *La Bibliofilia* 77 (1975), 227-51.

102 Falaschi, 223.

103 The most up-to-date list of Vavassore's maps is available from Rodney Shirley.

104 Del Badia, 24-30.

105 Neil Harris, Marin Sanudo, Forerunner of Melzi, Part I', *La Bibliofilia* 95 (1993), 1-37, especially 34-37, where he discusses the problems and magnitude of the loss of early editions hinted at only through ledgers, book lists, and legal documents.

106 Fabia Borroni, 'La pianta di Venezia del Guadagnino', *La Bibliofilia* 71 (1969), 265-68. Recent descriptions of such composite atlases include Isabella Zedda Macciò, 'Una "raccolta Lafreri" presso la Biblioteca Universitaria di Cagliari', In *La lettura geografica, il linguaggio geografico, i con-*

tenuti geografici a servizio dell'uomo, ed. Cosimo Palagiano and Emanuele Paratore, (Bologna: Pàtron Editore, 1991), 123–43; Lelio Pagani, 'Una raccolta di carte geografiche del secolo XVI', *Bergomum: Studi di storia arte e letteratura* (1979), 1–6; Albert Buvé, 'De italiaanse kartografie in de XVIe eeuw en de Mercatorverzameling', *Annalen, Oudheidkundige Kring van het Land van Waas* 75 (1972), 145–53; Rodney W. Shirley, 'Something Old, Something New in Grenoble', *Journal of the International Map Collectors' Society* (1992), 37–38, 40; Rodney W. Shirley, 'A Rare Italian Atlas at Hatfield House', *Map Collector* 60 (1992), 1–21; Sotheby Parke Bernet, 'Catalogue of Highly Important Maps and Atlases', 15 April 1980, 31–33; Dirk De Vries, 'Atlases and Maps from the Library of Isaac Vossius (1618–1689)', *International Yearbook of Cartography* 21 (1981), 177–93; Rudolf Kinauer, 'Neue Karten zum Atlas des Freiherrn Philipp von Stosch', In *Festschrift Josef Stummvoll*, ed. Josef Mayerhöfer and Walter Ritzer, (Vienna: Verlag Brüder Hollinek, 1970), 177–90.

107 Woodward, *Five Centuries*, 45–49.

108 Francesco Ehrle, *Roma al tempo di Giulio III: La pianta di Roma di Leonardo Bufalini del 1551* (Rome: Danesi, 1911).

109 For information on Pagano, see Leo Bagrow, *Matheo Pagano. A Venetian Cartographer of the 16th Century. A Descriptive List of his Maps* (Jenkintown, PA: George H. Beans Library, 1940); Beans, 'Vavassore and Pagano', 73; and Marcel Destombes, 'La grande carte d'Europe de Zuan Domenico Zorzi (1545) et l'activité cartographique de Matteo Pagano à Venise de 1538 à 1565', in *Etudes d'Histoire de la Géographie et de la Cartographie*, ed. Jósef Babicz, Monografie z Dziejow Nauki i Techniki 87 (Warsaw: 1973), 115–29.

110 Gastaldi, *Descrittione*, ff. A3 and A4.

111 Engraved by Pagano 20 Dec 15[38–39], impression in the University of Leiden Library. See Roberto Almagià, 'La più antica carta stampata del Piemonte', *L'Universo* 6 (1925), 985–89. The most up-to-date list of Pagano's maps is available from Rodney Shirley.

112 Manzi, 811–18. Hind, nos. 19–22, 170–71 and plates 734–35.

113 Adolf Erik Nordenskiöld, *Periplus: An Essay on the Early History of Charts and Sailing Directions* (Stockholm: 1897), 159.

114 The *World Encompassed: An Exhibition of the History of Maps held at the Baltimore Museum of Art* (Baltimore: Walters Art Gallery, 1952), item 62, plate XX.

115 Conor Fahy, 'The Venetian Ptolemy of 1548', In *The Italian Book 1465–1800: Studies Presented to Dennis E. Rhodes on his 70th Birthday*, ed. Denis V. Reidy, (London: The British Library, 1993), 89–115.

NOTES TO LECTURE II

1 Francesco Ehrle, *Roma prima di Sisto V: La pianta di Roma Du Pérac-Lafréry del 1577 riprodotta dall'esemplare esistente nel Museo Britannico. Contributo alla storia del commercio delle stampe a Roma nel secolo 16 e 17* (Rome: Danesi, 1908); Roberto Almagià, 'Nota su alcuni incisori e stampatori veneti e romani di carte geografiche', In *Carte geografiche a stampa di particolare pregio o rarità dei secoli XVI e XVII, esistenti nella Biblioteca Apostolica Vaticana*, ed. Roberto Almagià. (Vatican City: Biblioteca Apostolica Vaticana, 1948) II, 115-20; Fabia Borroni, *Carte, piante e stampe storiche delle raccolte lafreriane della Biblioteca Nazionale di Firenze* (Rome: Ministero per i beni culturali e ambientali, 1980); Richard Krautheimer, *Roma Alessandrina: the remapping of Rome under Alexander VII, 1655–1667* (Poughkeepsie: Vassar College, 1982).

2 Paolo Bellini, 'Stampatori e mercanti di stampe in Italia nei secoli XVI e XVII', *I quaderni del conoscitore di stampe* 26 (1975), 19-22; Maria Antonietta Bonaventura, 'L'industria e il commercio delle incisioni nella Roma del '500', *Studi romani: Rivista bimestrale dell'Istituto di Studi Romani* 8 (1960), 430-36; Maria Raffaella Caroselli, 'Commercio librario a Roma nel secolo XV', *Economia e Storia* 25 (1978), 221-37; Jacques Kuhnmünch, 'Le commerce de la gravure à Paris et à Rome au XVIIe siècle', *Nouvelles de l'estampe* 55 (1981), 6-17.

3 Maria Cristina Misiti, 'Antonio Salamanca: Qualche chiarimento biografico alla luce di un'indagine sulla presenza spagnola a Roma nel '500', In *La stampa in Italia nel Cinquecento in Rome*, edited by Marco Santoro, (Rome: Bulzoni Editore, 1992); David Landau and Peter Parshall, *The Renaissance Print: 1470–1550* (New Haven and London: Yale University Press, 1994).

4 Judith Hook, *The Sack of Rome* (London: 1972).

5 Antonino Bertolotti, *Artisti veneti in Roma nei secoli XV, XVI e XVII: Studi e ricerche negli archivi romani* (Bologna: Arnaldo Forni, 1884); P. S. Leicht, 'L'editore veneziano Michele Tramezino ed i suoi privilegi', In *Miscellanea di scritti di bibliografia ed erudizione in memoria di Luigi Ferrari*, (Florence: 1952), 357-67; Alberto Tinto, *Annali tipografici dei Tramezzino* (Florence: Leo S. Olschki, 1968).

6 Ehrle, 11-15; Antonino Bertolotti, *Artisti subalpini in Roma nei secoli XV, XVI e XVII: Ricerche e studi negli archivi romani* (Mantua: Mondovì, 1884); Antonino Bertolotti, *Artisti francesi in Roma nei secoli XV, XVI e XVII: Ricerche e studi negli archivi romani* (Mantua: G. Mondovi, 1886); F. Roland, 'Un Franc-Comtois éditeur et marchand d'estampes à Rome au XVIe siècle: Antoine Lafréry (1512-1577)', *Mémoires de la Société d'Émulation du Doubs* 5 (1910), 320-78.

7 Misiti, 560.

8 Christian Hülsen, Das 'Speculum Romanae Magnificentiae des Antonio Lafreri", In *Collectanea variae doctrinae Leoni S. Olschki*, (Munich: Rosenthal,

1921), 121–70; Hans P. Kraus, *Lafreri's Speculum Romanae Magnificentiae from the Library of the Earl of Crawford, the Most Extensive Copy Known* (New York: H. P. Kraus, 1948); Bates Lowry, 'Notes on the *Speculum Romanae Magnificentiae* and Related Publications', *Art Bulletin* 34 (1952), 46–50; Lawrence R. McGinniss, *Catalogue of the Earl of Crawford's 'Speculum romanae Magnificentiae' now in the Avery Architectural Library* (New York: Avery Architectural Library, Columbia University, 1976); Bernard Quaritch, *The 'Speculum Romanae Magnificentiae' of Antonio Lafreri Together with a Description of a Bertelli Collection of Maps* (London: Bernard Quaritch, 1928).

9 Borroni, *Carte, piante e stampe storiche*, describes the atlases in the Biblioteca Nazionale, Florence; the standard list of maps is still Ronald Vere Tooley, 'Maps in Italian Atlases of the Sixteenth Century, Being a Comparative List of the Italian Maps issued by Lafreri, Forlani, Duchetti, Bertelli and Others, Found in Atlases', *Imago Mundi* 3 (1939), 12–47.

10 Ehrle, 20; Gian Ludovico Masetti Zannini, *Stampatori e librai a Roma nella seconda metà del Cinquecento: Documenti inediti* (Rome: Fratelli Palombi Editori, 1980).

11 Gian Ludovico Masetti Zannini, 'Rivalità e lavoro di incisori nelle botteghe Lafréry-Duchet e de la Vacherie', In *Les fondations nationales dans la Rome pontificale*, (Rome: École Française de Rome, 1981), 547–66.

12 Horatio Brown, *The Venetian Printing Press: An Historical Study based upon Documents for the Most Part Hitherto Unpublished* (New York: G. P. Putnam, 1891); Richard MacKenney, *Tradesmen and Traders: The World of the Guilds in Venice and Europe, c. 1250–c. 1650* (London: Croom Helm, 1987).

13 MacKenney, 107.

14 These figures are taken from Tooley's catalogue, probably a significant underestimate.

15 David Woodward, 'Paolo Forlani: Compiler, Engraver, Printer, or Publisher?' *Imago Mundi* 44 (1992), 45–64.

16 David Woodward, *The Maps and Prints of Paolo Forlani: A Descriptive Bibliography* (Chicago: The Newberry Library, 1990).

17 Vittorio Zonca, *Novo teatro di machine et edificii per varie et sicure operationi. Co[n] le loro figure tagliate in rame e la dichiaratione, e dimostratione di ciascuna. Opera necesaria ad architetti, e a quelli, chi di tale studio si dilettano* (Padua: P. Bertelli, 1607).

18 Abraham Bosse, *Traicté des manières de graver en taille douce* (Paris, 1645). Furio de Denaro, ed., *Domenico Tempesti e I discorsi sopra l'intaglio ed ogni sorte d'intagliare in rame da lui provate e osservate dai più grand'huomini di tale professione*, (Florence: Studio Per Edizioni Scelte, 1994). Estimates of the time of inking are difficult to obtain. Factors include the complexity of the image and the dexterity of the worker. Even the extremely informative account of how British hydrographic charts were engraved and printed from copper does not provide the information. See Roy J. L. Cooney, Chart Engraving at the

Hydrographic Department, 1951–1981', *Cartographic Journal* 23, 2 (1986), 91–98. But a modern printmaker, Andrew Rubin of the Tandem Press, Madison, Wisconsin, has informed me that, although Tempesti's claim of 50 impressions a day is impressive, it would have been practicable to maintain inking a simple deeply cut copper plate of 29.5 × 43.4 cm. (*reale*) at the rate of 10–12 minutes each, particularly after the worker had become accustomed to the routine.

19 *Nova Reperta*. Engraved by Theodor Galle; Hollstein 7:424, p.87.

20 Annette Manick, 'A Note on Printing Inks', In *Italian Etchers of the Renaissance*, ed. Sue Welsh Reed and Richard Wallace (Boston: Museum of Fine Arts, 1989), xliv–xlvii.

21 De Denaro, 171.

22 De Denaro, 160.

23 De Denaro, 160.

24 Landau and Parshall, 31.

25 Catherine Wilkinson-Zerner, *Juan de Herrera: Architect to Philip II of Spain* (New Haven: Yale University Press, 1993).

26 Zonca, 78.

27 Landau and Parshall, 31.

28 The author plans such a study using the 1565 world map engraved by Paolo Forlani, for which states are recorded until the end of the sixteenth century. See Woodward, *Forlani Bibliography* 35.

29 Giuseppe Sergio Martini, *La bottega di un cartolaio fiorentino della seconda metà del quattrocento: Nuovi contributi biografici intorno a Gherardo e Monte di Giovanni* (Florence: Leo S. Olschki, 1956). In addition to paper, figures include that for an unworked goatskin (3s 4d), a sheepskin (4s), and a scraped calfskin (5s).

30 Conor Fahy, *Printing a Book at Verona in 1622* (Paris: Fondation Custodia, 1993).

31 The fullest source are payments recorded in the Chigi and Barberini households during the 1650s and 60s. See Consaagra, 400–401. Cardinal Francesco Barberini contracted the printing of the intaglio illustrations for Giorgio Pachimere's *Storia bizantina* and paid Antonio de Lanna 14.55 scudi for printing 4,846 sheets. Giovanni Giacomo de Rossi was paid 3 giulii per 100 sheets for 3,100 sheets, or a total 9.33 scudi. Earlier, Giovanni Giacomo had received 9.55 scudi from the same cardinal for printing 1,350 sheets. In 1666, the Chigi paid Pietro Paolo Giuliani 18.5 scudi for printing 12 images on silk 'with great diligence' as well as 200 *reale* sheets, which had been glued on the backings.

32 Robert Dossie, *The Handmaid to the Arts* (London: J. Nourse, 1758). Dossie reports that 'When the printer has done his work, he hangs on cords, that are clean and drawn tight, all the prints he has taken off . . . After which they must be taken down, and putting them together dozen by dozen into

the press, to take away the creasing of the cord, he leaves them there a day or two.' It would not have been practicable to tie up a rolling press in this way, so a standing press must have been meant.

33 Venice, Biblioteca Nazionale Marciana, Rari 138.c.4.

34 Jodoco Del Badia, 'La bottega di Alessandro di Francesco Rosselli merciaio stampatore', *Miscellanea fiorentina di erudizione e di storia* 2 (1894), 24–30.

35 Ehrle, appendix 9.

36 Consagra, various appendixes.

37 These figures are extrapolated from data provided by the accounts cited earlier in this section and the corpus of the engraver Paolo Forlani. For the prices of Italian maps several centuries later, see Calcografia Camerale, *Catalogo delle stampe tratte dai rami intagliati a bullino, ed in acqua forte di proprietà della Calcografia Camerale aumentato con l'acquisto dell'intera Calcografia Volpato e di molti altri rami di opere celebri vendibili in Via del Corso, n. 139 a prezzi fissi* (Rome: Vincenzo Poggioli, 1826).

38 David Woodward, 'The Evidence of Offsets in Renaissance Italian Maps and Prints', *Print Quarterly* 8 (1991), 235–51.

39 David Rosand and Michelangelo Muraro, *Titian and the Venetian Woodcut*, (Washington: International Exhibitions Foundation, 1976). The catalogue description does not mention the offset; I owe this discovery to David Landau.

40 Lucia Nuti, 'The Perspective Plan in the Sixteenth Century: The Invention of a Representational Language', *Art Bulletin* 76 (1994), 105–28.

41 David Woodward, 'The Forlani Map of North America', *Imago Mundi* 46 (1994), 29–40. See 'WATERMARKS' below for a fuller analysis of the use of this evidence.

42 Woodward, Offsets', 237, where the print is reproduced.

43 Landau and Parshall, 300.

44 Karen Cook, Map Library, British Library, pointed out to me that the maps are *printed*, not pasted, on the back of the prints, as has previously been reported.

45 Curt F. Bühler, Last Words on Watermarks,' *Papers of the Bibliographical Society of America* 67 (1973), 1–16. For a rather more optimistic, but nevertheless appropriately cautious, evaluation of the value of watermarks for the print historian, see Elizabeth Lunning, 'Characteristics of Italian Paper in the Seventeenth Century', In *Italian Etchers of the Renaissance*, ed. Sue Welsh Reed and Richard Wallace, (Boston: Museum of Fine Arts, 1989), xxxii–xliii.

46 David Woodward, *Catalogue of Watermarks in Italian Maps, ca. 1540–1600* (Florence: Leo S. Olschki, 1996).

47 David Woodward, 'The Correlation of Watermark and Paper Chemistry in Sixteenth Century Italian Printed Maps', *Imago Mundi* 42 (1990), 84–93.

48 David Woodward, 'The Analysis of Paper and Ink in Early Maps:

Opportunities and Realities', In *Essays in Paper Analysis*, ed. Stephen Spector, (Washington D.C.: Folger Books, 1987), 200–21.

49 David Woodward, 'The Sixteenth-Century Italian Composite Atlas', In *Images of the World: A History of the Atlas*, ed. John A. Wolter and Ronald Grim, (Washington, D.C.: Library of Congress, 1996).

50 David Woodward, 'New Tools for the Study of Watermarks on Sixteenth-Century Italian Printed Maps: Beta Radiography and Scanning Densitometry', In *Imago et Mensura Mundi: Atti del IX Congresso Internazionale di Storia della Cartografia*, ed. Carla Clivio Marzoli (Rome: Istituto della Enciclopedia Italiana, 1985) Vol. 2, 541–52.

51 Albert Ganado, 'Description of an Early Venetian Sixteenth Century Collection of Maps at the Casanatense Library in Rome', *Imago Mundi* 34 (1982),

52 The Marciana atlas offers an example of the ravages these Italian composite atlases have undergone. Once an atlas has reached the haven of a library, its contents might be thought to be safe. Such is not the case with the Marciana atlas. Rodolfo Gallo first described the atlas in 1954 and listed the maps: Rodolfo Gallo, *Carte geografiche cinquecentesche a stampa della Biblioteca Marciana e della Biblioteca del Museo Correr di Venezia* (Venice: 1954). I examined it on 21 July 1978 and four maps had already been removed. On 11 March 1995 I examined it again, and noted that some of its even more valuable maps had disappeared: 1. Universale descrittione (the Forlani-Bertelli 1565 map of the world), 2. Mappamondo cuoriforme (the Cimerlino map of the world), 3. Mappamondo in due emisferi (the Floriano map of the world) and 4. Disegno del discoperto della Nova Francia (the Forlani map of North America), and replaced (rather pathetically) with two 18th-century maps. The stubs of the missing maps are still visible. The bibliographical evidence lost by this act of vandalism to the earliest Italian composite atlas is incalculable.

53 The Esecutori contra la Bestemmia were given control of printing in 1543. See Paul Grendler, *The Roman Inquisition and the Venetian Press, 1540–1605* (Princeton: Princeton University Press, 1977), 26 and Richard J. Agee, 'The Privilege and Venetian Music Printing in the Sixteenth Century', Ph.D Dissertation, Princeton University, 1982. For other sources of information on the control of printing, see Leonardus Vytautas Gerulaitis, *Printing and Publishing in Fifteenth-Century Venice* (Chicago: American Library Association, 1976); Rinaldo Fulin, 'Primi privilegi di stampa in Venezia', *Archivio Veneto* 1 (1871), 160–64; Maria Grazia Blasio, 'Privilegi e licenze di stampa a Roma fra Quattro e Cinquecento', *La Bibliofilia* 90 (1988), 147–59; Venice, Biblioteca Nazionale Marciana, manuscript by Horatio F. Brown, 'Privilegi veneziani per la stampa (1527–1597) ca. 1890; Carlo Castellani, *I privilegi di stampa e la proprietà letteraria in Venezia* (Venice: 1888); and Pierina Fontana, 'Inizi della proprietà letteraria nello Stato Pontificio (Saggio di

documenti dell'Archivio Vaticano)', *Accademie e biblioteche d'Italia* 3 (1929–30), 204–21.

54 Venice, Archivio di Stato di Venezia, Giustizia Vecchia, busta 49 reg. 79 (lettera P, c. 68v). See Giacomo Moro, 'Insegne librarie e marche tipografiche in un registro veneziano del '500', *La Bibliofilia* 91 (1989), 51–80.

55 Rodolfo Gallo, 'Gioan Francesco Camocio and his large Map of Europe', *Imago Mundi* (1950), 93–102. For background on the enforcing of the Counter Reformation's ban on prohibited books, involving several publishers who were also printmakers, see Paul F. Grendler, *The Roman Inquisition and the Venetian Press, 1540–1605* (Princeton: Princeton University Press, 1977), 162–69.

56 See Henri Zerner, L'estampe érotique au temps de Titien', in *Tiziano e Venezia: Convegno Internazionale di Studi, Venezia 1976* (Vicenza: Neri Pozza Editore, 1980), 85–90. For photographic reproductions of the only complete copy of *I modi*, see P. Aretino, *Poesie varie*, ed. G. Aquilecchia and A. Romano (Rome: Salerno Editrice, 1992), 103–14, and Lynne Lawner, ed., *I Modi: The Sixteen Pleasures* (Evanston: Northwestern University Press, 1988).

57 Rodolfo Gallo, 'Antonio Florian and his Mappemonde', *Imago Mundi* 6 (1949), 35–38.

58 Consagra, 88.

59 Agee, 9.

60 Consagra, 107.

61 Gerulaitis, 33–35.

62 Venice, Archivio di Stato di Venezia, Senato Terra, Filza 22. Supplica e privilegio.

63 Brown, *Venetian Printing Press*, 102.

64 Venice, Archivio di Stato di Venezia, Senato Terra 46, fol. 10, 5 December 1567. See Roberto Almagià, *Monumenta Cartographica Vaticana* (Vatican City: Biblioteca Apostolica Vaticana, 1944–52), II, 116.

65 Paolo Forlani, *La prima parte della descrittione del Regno di Polonia*, 1568 (Tooley 454; Woodward, *Forlani Bibliography* 80) and *Il uero disegno della seconda parte dil Regno di Polonia*, 1568 (Tooley 455; Woodward, *Forlani Bibliography* 80).

66 Venice, Archivio di Stato di Venezia, Paolo da Canal. For further background on this early period of Venetian map publishing, see Lilian Armstrong, Benedetto Bordon, *Miniator,* and Cartography in Early Sixteenth-Century Venice', *Imago Mundi* 48 (1996), 65–92.

67 Landau and Parshall, 301.

68 Consagra, 82.

69 Forlani's only association with a privilege application was with Gastaldi on 11 September 1564 for his map of Asia Minor, no doubt *Il disegno d' Geografia moderna della prouincia di Natolia, et Caramania* . . . Gastaldi must

have given Forlani permission to engrave the map and put the words 'cum privilegio' on it, but in other cases no such formal arrangement seems to have been made. Forlani felt free to use the name of the 'excellent cosmographer' even while he was alive without further formalities. We do not know that he paid Gastaldi a royalty for using his compilations, nor whether they had frequent contact.

70 Paolo Forlani's patrons included Piero Badoero, Luigi Balbi, Daniele Banda, Nicolao Bandae IV, Cosimo Bartoli, Daniele Barbaro, Lodovico del Bene, Ottavio del Bene, Geronimo Canossa, Gio. Pietro Contarini, Alfonso dal Corno, Daniele Dana, Antonio Diedo, Jodoco á Meggen, Franco Morandi, Girolamo Murari, Andrea degli Orefici, Marcantonio Radici, Nicola Rambaldo, Tommaso da Ravenna, Alessandro Serego, Marco del Sole, Nicolaus Stopius, Antonio Tognale , Giovan Battista della Torre, Agostino Valier, Giovanni Battista Veculiano, Carlo Vicentino, Paolo Michaeli Vincentino, and Bartolommeo Zacco.

71 Paolo Forlani, *Universale descrittione di tutta la terra conosciuta fin qui* (Venice: Al segno del pozzo, 1562) (Tooley 9; Woodward, *Forlani Bibliography* 9.01). 'Impero che essendo ella di cosi nobile et honorata famiglia come è, son certo, che accettara la prottetione dell'opera, contra i mal dicenti, iquali altro non sa fare, che biasimare le altrui operatione mosse da buona et utile intentione, . .'

72 Paolo Forlani, *La descrittione di tutto il peru* (Venice, ca. 1565) (Tooley 93; Woodward, *Forlani Bibliography* 11).

73 G. Astegiano, 'La vita e le opere di Tommaso da Ravenna', *Bollettino del Museo Civico di Padova* n.s 1, 18 (1925), 49-70 and 236-60.

74 The plant is probably *lignum vitae*, known in Hispaniolan as *huiacum* or by the natives of Puerto Rico as *Hoaxacan*. See Robert S. Munger, Guaiacum, the Holy Wood from the New World', *Journal of the History of Medicine and Allied Sciences* 4 (1949), 196-229. I owe this reference to John Scarborough.

75 Mattioli was also the translator of Gastaldi's 1548 edition of Ptolemy's *Geography*. See Conor Fahy, 'The Venetian Ptolemy of 1548', In *The Italian Book 1465–1800: Studies Presented to Dennis E. Rhodes on His 70th Birthday*, ed. Denis V. Reidy, (London: The British Library, 1993), 89-115.

76 For useful gateways to the northern equivalent of this subject from the print historian's perspective, see Timothy Riggs and Larry Silver, *Graven Images, the Rise of Professional Printmakers in Antwerp and Haarlem* (Evanston, Illinois: Mary and Leigh Block Gallery, Northwestern University, 1993); William W. Robinson, '"This Passion for Prints": Collecting and Connoisseurship in Northern Europe during the Seventeenth Century', In *Printmaking in the Age of Rembrandt*, ed. Clifford S. Ackley, (Boston: Museum of Fine Arts, 1981), xxvii–xlviii; Nadine Orenstein, *Hendrik Hondius and the Business of Prints in Seventeenth-Century Holland* (Roosendall: Van Poll Publishers, 1995).

NOTES TO LECTURE III

1 Grant McCracken, *Culture and Consumption: New Approaches to the Symbolic Character of Consumer Goods and Activities* (Bloomington: Indiana University Press, 1988). The idea of the growing importance of consumer goods in a capitalist society was pioneered by Fernand Braudel, *The Structures of Everyday Life: The Limits of the Possible*, Vol. 1, *Civilization and Capitalism, 15th–18th Century* (New York: Harper & Row, 1981). See also Richard A. Goldthwaite, *Wealth and the Demand for Art in Italy, 1300–1600* (Baltimore and London: Johns Hopkins University Press, 1933).

2 Richard A. Goldthwaite, *The Building of Renaissance Florence: An Economic and Social History* (Baltimore and London: 1980).

3 Goldthwaite, Florence 407.

4 Michael Bury, 'The Taste for Prints in Italy to c. 1600', *Print Quarterly* 2 (1985), 12-26.

5 David Landau, 'Vasari, Prints and Prejudice', *Oxford Art Journal* 6, no. 1 (1983), 3-10.

6 This debate has centred on whether the use of maps in political and social administration was 'normal' in Italy before the third quarter of the sixteenth century. See John Marino, 'Administrative Mapping in the Italian States', *In Monarchs, Ministers, and Maps*, ed. David Buisseret. 5-25. (Chicago: University of Chicago Press, 1992). Marino's statement that 'mapping as a normal administrative way of looking at the world dates only from the third quarter of the sixteenth century' may need some qualification, or at least some discussion of the meaning of 'normal.' As more and more explicit references to the use of maps appear, however, it is more difficult to 'argue from silence'.

7 Chandra Mukerji, *From Graven Images: Patterns of Modern Materialism* (New York: Columbia University Press, 1983), 37.

8 Mukerji, 97-98.

9 Mukerji, 81.

10 W. F. Ganong, *Crucial Maps in the Early Cartography and Place Nomenclature of the Atlantic Coast of Canada* (Toronto: University of Toronto Press for the Royal Society of Canada, 1964), 264.

11 Paolo Forlani, *La descrittione dell'Europa, et parte dell'Africa, et dell'Asia, seco[n]do l'uso de nauiga[n]ti del S. Giac.*^{mo} *Home[m] portughese*, 1569; Ronald Vere Tooley, 'Maps in Italian Atlases of the Sixteenth Century, Being a Comparative List of the Italian Maps issued by Lafreri, Forlani, Duchetti, Bertelli and Others, Found in Atlases', *Imago Mundi* 3 (1939), 12-47, map 34, David Woodward, *The Maps and Prints of Paolo Forlani: A Descriptive Bibliography* (Chicago: The Newberry Library, 1990), map 81.01.

12 The many painted maps that adorned the walls of the Vatican, the Palazzo Vecchio in Florence, the Ducal Palace in Venice, and many other villas,

monasteries, and palaces are not our focus here. These have attracted a substantial and now rapidly growing literature. See Roberto Almagià, *Monumenta Cartographica Vaticana* (Vatican City: Biblioteca Apostolica Vaticana, 1944-52) and the recent facsimile, *La Galleria delle carte geografiche in Vaticano* (Modena: F. C. Panini, 1994); Loren Partridge, 'The Room of Maps at Caprarola, 1573-1575', *Art Bulletin* 77 (1995), 413-44; Juergen Schulz, 'Maps as Metaphors: Mural Map Cycles in the Italian Renaissance', In *Art and Cartography: Six Historical Essays*, ed. David Woodward, (Chicago: University of Chicago Press, 1987), 97-122. More examples in smaller villas have yet to be described, however: an example is the series of sixteenth-century maps based on Ortelius in a loggia of the Villa now occupied by Giunti Editore in Florence. Smaller painted or manuscript maps were also framed for display. For example, in 1418, Cosimo de' Medici had 'a beautiful world map' hanging in his office. In 1488 Luigi d'Ugolino Martelli, a prominent merchant, had eight painted maps worth 9 écu (or 9 large florins). In 1497 in the house of Giovanni di Francesco Tornabuoni, there were two world maps and a painted map of Spain, Burgundy, and France in gilded frames. In 1513, Giovanni di Francesco Inghirami owned painted maps of Italy, Florence, and Naples. See John Kent Lydecker, 'The Domestic Setting of the Arts in Renaissance Florence', Ph.D, Johns Hopkins, 1987, 62-63, 68, 130. Spallanzani and Bertelà report the following cartographic items inventoried at the death of Lorenzo de' Medici 23 December 1512 by commission of Lorenzo di Piero de' Medici:

Nella chamera di detta sala [=sala grande nella loggia terrena]
Uno quadro dipintovi una Italia f[lorins]25
Uno quadro di legno dipintovi la Spagna f12

Chamera delle dua letta
Uno tondo grande dipintovi uno universale chon uno festone di nocie intagliato intorno, diametro br.— f50 (This could be a globe or circular decoration with a carved walnut garland).

In Camera della sala grande
Un colmo di br 3 lunghio, dipintovi una Terrasanta f6
Un colmo di br 2 1/2, dipintovi la Spagna f8

Nell'antichamera del cassone
Un colmo di br 1 1/2 dipintovi una Roma f20
Un colmo di br 4 1/2 dipintovi l'universo f50
Un colmo di br iii 1/2 dipintovi una Italia f25

Nello scrittoio (no florin values entered)
Una carta dipintavi Italia
Un'altra carta dipintavi il chastel di Milano
Una dipintavi el mappamondo
Uno dipintavi l'apamondo
Una carta dipintavi tre reami d'India

Un'altra dipintovi l'apamondo
Un'altra dipintovi Italia
Una carta da navichare
Un'altra dentrovi l'apamondo
Una carta stretta e lungha dentrovi più paesi
Una carta dentrovi Roma
Uno colmo di br 1 1/2 dipintovi una Roma f20
Uno colmo di br 4 1/2 dipintovi l'universo
Florence, Archivio di Stato di Firenze, Mediceo Avanti Principato, Filza CLXV. Marco Spallanzani and Giovanna Gaeta Bertelà, *Libro d'inventario dei beni di Lorenzo il Magnifico* (Florence: Associazione 'Amici del Bargello' (Studio Per Edizioni Scelte), 1992).

13 Francis Haskell, *Patrons and Painters: A Study in the Relations Between Italian Art and Society in the Age of the Baroque* (New Haven: Yale University Press, 1980); Marilyn Aronberg Lavin, *Seventeenth-Century Barberini Documents and Inventories of Art* (New York: New York University Press, 1975).

14 Lydecker, 16–17.

15 Venetian inventories are found in various sources of the Archivio di Stato di Venezia, including the *Notarile, Miscellanea notai diversi* and *the Giudici di Petizion.*

16 Isabella Palumbo-Fossati, 'L'interno della casa dell'artigianato e dell'artista nella Venezia del Cinquecento', *Studi Veneziani* 8 (1984), 109–53; Marco Ferro, *Dizionario del diritto comune e veneto* (Venice: A. Santini e figlio, 1845–47).

17 V. G. Ménage, 'The Map of Hajji Ahmed and Its Makers', *Bulletin of the School of Oriental and African Studies: University of London* 21 (1958), 291–314.

18 George Kish, *The Suppressed Turkish Map of 1560* (Ann Arbor: William L. Clements Library, 1957). The term 'in balla' is problematic, as the Hajji Ahmed map is on a cordiform projection and is not a globe. However, the term does not always refer to a globe but can also refer to a map on a projection that appears to mimic the spherical shape of the earth.

19 Federica Ambrosini, '"Descrittioni del Mondo" nelle case venete dei secoli XVI e XVII', *Archivio Veneto* 5 (1981), 67–79, esp. 70. The function of such displays was no doubt partially to set their owners apart from their less cultured neighbors, a concept explored in Pierre Bourdieu, *Distinction: A Social Critique of the Judgement of Taste* (London: Routledge, 1984).

20 Palumbo-Fossati, 147.

21 Venice, Biblioteca Nazionale Marziana, ms Ital IV 133 (=5103).

22 Ambrosini, 74.

23 M. T. Cipollato, 'L'eredità di Federico Contarini: gli inventari della collezione e degli oggetti domestici', *Bollettino dell'Istituto di storia della società e dello Stato veneziano* 3 (1961), 228 and 232.

24 Ambrosini, 76 and 72.

25 Sometimes the inventories tell us that the maps were neglected, the vestige of a geographical interest of the former proprietor not shared by the heirs. For several examples of damaged or old maps, or those relegated to storage (for example, the one that Anna Trevisan threw out in 1624), see Ambrosini, 70, fn8.

26 Juergen Schulz, 'Jacopo de' Barbari's View of Venice: Map Making, City Views, and Moralized Geography before the Year 1500', *Art Bulletin* 60 (1978), 425-74.

27 Ambrosini, 75.

28 Francesca Consagra, 'The De Rossi Family Print Publishing Shop: A Study in the History of the Print Industry in Seventeenth-Century Rome', Ph.D. Dissertation, The Johns Hopkins University, 1993, 347.

29 Ambrosini, 76.

30 Simona Savini Branca, *Il collezionismo veneziano nel '600* (Padua: CEDAM, 1965).

31 See Frederik Caspar Wieder, 'Lafreri versus Ortelius', In *Nederlandsche Historisch-Geografische Documenten in Spanje*, (Leiden: E. J. Brill, 1915), 111-58; Roberto Almagià, 'La diffusion des produits cartographiques flamands en Italie au XVIe siècle', *Archives internationales d'histoire des sciences* 7 (1954), 46-48; Günter Schilder, 'The Cartographical Relationships between Italy and the Low Countries in the Sixteenth Century', In *Imago et Mensura Mundi: Atti del IX Congresso Internazionale di Storia della Cartografia*, ed. Carla Clivio Marzoli (Rome: Istituto della Enciclopedia Italiana, 1985), 2 vols., I: 265-78. See also Schilder's ongoing *Monumenta Cartographica Neerlandica* (Alphen aan den Rijn: Canaletto, 1986-), for references to Dutch influence on Italian maps, and vice-versa.

32 See the description of this genre of the *conclusione* in Consagra, 184-86. I would also like to acknowledge the generous and enthusiastic help of Lucia Nuti, Louise Rice, and John Beldon Scott in compiling this section.

33 F. W. H. Hollstein, *German Engravings, Etchings, and Woodcuts, ca. 1400–1700* (Amsterdam: M. Hertzberger, 1954-).

34 The arms of Francesco and Antonio are similar, both including a Maltese cross in addition to the familiar three bees, but Antonio's usually have the cross in the background, while in Francesco's the cross is at the top, as here.

35 Cesare Ripa, *Iconologia overo Descrittione dell' Imagini universali cavate dall'antichita et da altri luoghi Da Cesare Ripa Perugino. Opera non meno utile, che necessaria a Poeti, Pittori, & Scultori, per rappresentare le virtu, vitij, affetti, & passioni humane* (Rome: Heredi di Gio. Gigliotti, 1593).

36 Several instances of this link come to mind: in Chinese cartography, there is the famous story of Lady Zhao, 'the wonder of needlework', who embroidered a map for the king of Wu in the third century AD. See Cordell Yee, Chinese Cartography among the Arts', 140-42. Zhang Heng, a Chinese astronomer is reported to have 'cast a net around heaven and earth and cal-

culated on the basis of it.' See Yee, Taking the World's Measure', 125; both references in J. B. Harley and David Woodward, eds., *The History of Cartography*, (Chicago: University of Chicago Press, 1994), Vol. 2.2. The idea is echoed in John Donne: 'For of Meridians, and Parallels, Man hath weav'd out a net, and this net throwne Upon the Heavens, and now they are his owne.' *An Anatomy of the World: The First Anniversary.*

37 Eva Schulz, 'Notes on the History of Collecting and of Museums in the Light of Selected Literature of the Sixteenth to the Eighteenth Century', *Journal of the History of Collections* 2 (1990), 205–18.

38 Bury, 26.

39 Peter W. Parshall, 'The Print Collection of Ferdinand, Archduke of Tyrol', *Jahrbuch der kunsthistorischen Sammlungen in Wien* 78 (1982), 139–84, esp. 180–81.

40 Roberto Almagià, 'Nota su alcuni incisori e stampatori veneti e romani di carte geografiche', In *Carte geografiche a stampa di particolare pregio o rarità dei secoli XVI e XVII, esistenti nella Biblioteca Apostolica Vaticana* (Vatican City: Biblioteca Apostolica Vaticana, 1948), II: 115–20.

41 Antonio Lafreri, *Indice delle tavole moderne di geografia della maggior parte del mondo di diversi auttor* (Rome: Antoine Lafréry, ca. 1573), unique copy in the Biblioteca Marucelliana, Florence. See also Fabia Borroni, Carte, *piante e stampe storiche delle raccolte lafreriane della Biblioteca Nazionale di Firenze* (Rome: Ministero per i beni culturali e ambientali, 1980), who correctly points out (xxxix) that the pamphlet cannot have been published in 1572 as has frequently been estimated, because it contains a map that was published in 1573.

42 Tooley 35.

43 Tooley 84–92, 93, and 80, respectively.

44 Alberto Tinto, *Annali tipografici dei Tramezzino* (Florence: Leo S. Olschki, 1968), xiv, fn.3.

45 Borroni, xxxvii.

46 Jan H. Hessels, *Abraham Ortelii . . . Epistulae* (Cambridge, 1887, reprinted Osnabrück: Zeller, 1969), 368 (The map is Tooley 38); Roberto Almagià, 'Un manipolo di preziose carte a stampa di regioni italiane dei secoli XVI e XVII', *L'Universo* 4 (1923), 361–71.

47 Paula Findlen, *Possessing Nature: Museums, Collecting, and Scientific Culture in Early Modern Italy* (Berkeley: University of California Press, 1994), 112. See also Oliver Impey and Arthur MacGregor, eds., *The Origins of Museums: The Cabinet of Curiosities in Sixteenth- and Seventeenth-century Europe* (Oxford: Clarendon Press, 1985).

48 Lawrence Paul Rose, 'Jacomo Contarini (1536–1595), a Venetian Patron and Collector of Mathematical Instruments and Books', *Physis* 18 (1976), 117–30.

49 G. Astegiano, 'La vita e le opere di Tommaso da Ravenna', *Bollettino del Museo Civico di Padova* n.s. 1, 18 (1925), 49–70 and 236–60.

50 Cristiana Scappini, Maria Pia Torricelli, and Sandra Tugnoli Pattaro, eds., *Lo studio Aldrovandi in Palazzo Pubblico (1617–1742)* (Bologna: Editrice Clueb, 1993).

51 Schulz (Eva), Collecting', 206.

52 Samuel Quicchelberg, *Inscriptiones vel tituli theatri amplissimi, complectentis rerum universitatis singulas materias et imagines eximias, ut idem recte quoque dici possit* (Munich: Adam Berg, 1565).

53 Elizabeth M. Hajós, 'The Concept of an Engravings Collection in the Year 1565: Quicchelberg, *Inscriptiones vel tituli theatri amplissimi*', *Art Bulletin* 40 (1958), 152–53.

54 Schulz (Eva), Collecting', 209.

55 Translation from David Landau and Peter Parshall, *The Renaissance Print: 1470–1550* (New Haven and London: Yale University Press, 1994), 359. A useful sample of images of broadsheet and print sellers is found in Sigfred Taubert, *Bibliopola: Bilder und Texte aus der Welt des Buchhandels* (Hamburg: Ernst Hauswedell, 1966). See also Alfredo Petrucci, 'Aspetti della vecchia Roma. I venditori ambulanti in una stampa antica', *Capitolium* (1932).

56 Maps were sold along with popular satirical prints, such as 'The World Upside Down' of the 'Land of Cockaigne.' See Giuseppe Cocchiara, *Il Paese di Cuccagna e altri studi di folklore* (Turin: Einaudi, 1956), and *Il mondo alla rovescia* (Turin: Paolo Boringhieri, 1981). See the entry on Nicccolo Nelli by David Acton in Sue Welsh Reed and Richard Wallace, *Italian Etchers of the Renaissance & Baroque* (Boston: Museum of Fine Arts, 1989), 47–53. For a general discussion of the 'world upside down' theme, see Roger Chartier, ed., *Cultural History: Between Practices and Representations*, (Ithaca: Cornell University Press, 1988), 116–26. For the history of popular prints in general, see Achille Bertarelli, *L'Imagerie Populaire Italienne* (Paris: Duchartre & van Buggenhoudt, 1929); Anna Omodeo, 'Mostra di stampe popolari venete del '500', In *Gabinetto Disegni e Stampe degli Uffizi* 20 (Florence: Leo S. Olschki, 1965); Anna Maria Petrioli Tofani, 'Stampe popolari venete del Cinquecento agli Uffizi', *Antichità viva* 4 (1965), 62–71; and P. Toschi, *Stampe popolari italiane* (Milan: 1964). The reporting of the wars with the Turks was a favorite subject, and allegorical prints commented on the closely followed events. In one, the Pope, in the guise of St. Peter, and with the blessing of God, is trawling for Turks at the Battle of Lepanto (1571).

57 The annotation is on the 'Secunda Africae Tabula', Amsterdam, University Library, Map Room.

58 I would like to thank Dr. Albert Ganado for his invaluable assistance with this section. For further information, see the definitive two-volume work by Albert Ganado and Maurice Agius-Adalà, *A Study in Depth of 143 Maps*

Representing the Great Siege of Malta of 1565 (Valletta: Publishers Enterprises Group, 1994–96), 2 vols.

59 Ganado and Agius-Adalà, *Siege of Malta*, xvi.

60 J. B. Harley, 'Maps, Knowledge and Power', In *The Iconography of Landscape: Essays on the Symbolic Representation, Design and Use of Past Environments*, ed. Denis Cosgrove and Stephen Daniels, (Cambridge: Cambridge University Press, 1988), 277–312.

Appendix

Currencies quoted frequently in these lectures are the Venetian *ducato* (ducat), the Florentine *fiorino* (florin, replaced by the *scudo* in 1530), and the Roman *scudo,* all gold coins weighing about 3.5 grams. Although it is difficult to calculate equivalents of the currencies of the various Italian cities was in the sixteenth century, the Venetian gold *ducato* (also known as the *zecchino*) was very approximately equivalent to one Florentine or Roman *scudo* at the end of the sixteenth century.

Given their high intrinsic value, gold coins were used only in very large transactions, notably in international trade; everyday transactions were handled with small silver or silver-and-copper coins. Nearly all prices and wages were expressed in the *lira, soldo and denaro (£sd);* the lira itself was never coined, while some *soldi* and *denari* were. There were thus in existence two parallel currency systems. While the large coins in the first system remained stable over several centuries, the small ones in the second system did not, mainly because they were often debased by governments in need of revenue. The outcome was that the rate of exchange between the gold currency and the petty silver currency fluctuated over time—an outcome that could be exacerbated by changes in the price ratio of silver to gold.

The value of the *fiorino, ducato,* or *scudo* to the *lira* (*£*) was determined by the authorities in each city. In the middle of the thirteenth century, the gold florin was equivalent to £1 in Florence, but by 1534 it was equivalent to £7 10s. The exchange rate of the Venetian gold ducat to the lira increased from £6 4s in 1518 to £8 in 1562 to £10 in 1594 to £15 in 1638 as the petty currency became increasingly debased. For a summary of these problems of currency equivalents, with excellent bibliographies of sources of information on currencies in Italian cities in the Middle Ages and Renaissance, see Peter Spufford, *Handbook of Medieval Exchange,* Royal Historical Society Guides and Handbooks 13 (London: Royal Historical Society, 1986) and Carlo M. Cipolla, *Money in Sixteenth-Century Florence* (Berkeley: University of California Press, 1989).

1 *lira* (£) = 20 *soldi*
1 *soldo* = 12 *denari*
1 *scudo* = 100 *baiocchi*
1 *scudo* = 10 *giulii*
1 *quattrino* = 20 *baiocchi*
1 *giulio* = 10 *baiocchi*